The Fine Structure of Cotton

An Atlas of Cotton Microscopy

FIBER SCIENCE SERIES

Series Editor

L. REBENFELD

Textile Research Institute
Princeton, New Jersey

Other Volumes in Preparation

The Fine Structure of Cotton
An Atlas of Cotton Microscopy

Inez V. deGruy   Jarrell H. Carra   Wilton R. Goynes

Edited and with an Introduction by
Robert T. O'Connor

Southern Regional Research Center
New Orleans, Louisiana

MARCEL DEKKER, INC. New York   1973

This Atlas is dedicated to
Mary Lockett Rollins
with admiration and gratitude for sharing with her colleagues
her scientific knowledge of cotton as well as her personal friendship.

CONTENTS

PREFACE

This atlas presents a record of cotton fiber microscopy with brief descriptions of techniques used to obtain the photographs. It is, above all, a book for the industrial producer and the entire cotton industry, including the textile researcher whose own work is the basis for this review.

Cotton is the most important vegetable fiber used in spinning. It is a plant of the genus *Gossypium,* a member of the Malvaceae or mallow family, and is widely grown in tropical and subtropical regions all over the world. The seed hairs of cotton form the raw material for a large proportion of the world's textiles.

The date of the earliest cultivation of cotton is not actually known, but archeological evidence indicates that it was used in cloth as early as 3000 B.C. [1]. Reference to its cultivation in India was made in 350 B.C., but there is nothing to prove it was indigenous even to that country. Centuries before Christ, various travelers to India reported the existence of a plant bearing "wool" of a softness incomparable to that of any fiber they had known. During the seventh century it was introduced into China, but the Chinese used it as a garden shrub until about the year 1000 when the plant was cultivated for commercial use. Although it was brought to Japan in the eighth century, its growth was discontinued until the seventeenth century, i.e., at about the time the first attempt at its cultivation was made in the American colonies. It is reported that the first American cotton mill was built in 1787 at Beverly, Massachusetts, and in 1794 a patent was granted to Eli Whitney for his cotton gin.

In order of importance, the chief cotton-growing countries of the world are the United States, Russia, China, India, Brazil, and Egypt. Other areas of the world producing cotton are Pakistan, Turkey, Mexico, and Sudan.

The average cotton plant is a herbaceous shrub having a normal height of about 4–6 ft, although there are tree varieties in which the plant reaches a maximum of 15 or 20 ft. The most important botanical species included in the genus, *Gossypium*, are *hirsutum*, *barbadense*, *arboreum*, and *herbaceum*.

Gossypium hirsutum, of Central American origin, is a shrubby plant which reaches a maximum height of 6 ft. It is the species containing the Upland varieties which constitute 87% of world commerce in cotton. It is probably the origin of the green-seeded cotton now grown so extensively in the southern states.

Gossypium barbadense, believed to have originated in Peru, grows from 6–15 ft, was introduced into Egypt from America early in the eighteenth century, and is cultivated along the Nile. It is the extra-fine, extra-long-staple group of cottons which include Sea Island, the Egyptian Giza strains, Pima, and other American–Egyptian cottons. *G. barbadense*, which grows from black seeds, constitutes about 8% of the world's crop.

Gossypium arboreum includes the tree cottons and some of the native cottons of India and Pakistan. It grows as tall as 15–20 ft and includes both Asiatic and Indian cotton varieties. Its seeds are covered with a greenish fur.

Gossypium herbaceum averages about 4–6 ft and is grown from a seed encased in a gray down. Cottons from this species include the very short-stapled Asiatics, some Chinese, and most native Indian varieties. *G. herbaceum* and *G. arboreum* together constitute 5% of the world's cotton production.

The best growth conditions for cotton include a warm (60°–80° F av temp) climate and fairly moist, "loamy" soil. Under normal climatic conditions, cotton seeds germinate in 7–10 days after planting; in 35–45 days, flower buds known as squares appear. Open flowers are produced by the squares in 21–25 days, and 45–90 days after flowering, mature open bolls appear. The durations of these periods depend both on the variety of cotton and the environmental conditions.

A cotton boll contains 3–5 divisions called locks, each of which normally has about 9 seeds surrounded by fuzz fibers and a thick mass of cotton hairs or lint. Fuzz fibers are short hairs forming a "velvet-like" covering over the seed; they are morphologically distinct from the lint hairs and are easily discernible after the mature fibers have been removed. They differ from lint hairs by being shorter and coarser, but they are often indistinguishable from short lint fibers.

When the flower on the cotton plant opens, the fibers, which are single cells, begin to grow on the surface of the seed. There may be as many as 10,000 fibers on the seed—more than 250,000 to the boll. In the unopened boll the fiber attains its maximum length in 16–17 days, during which time the lumen, or central canal, is large and filled with plant juices from the protoplasm of the cell. Subsequently, the fiber begins to deposit secondary-wall cellulose diurnally in concentric layers from the outside or primary-wall toward the lumen. The primary wall is composed of cellulose, fats, waxes, pectic materials, and proteinaceous matter, while the secondary wall, the body of the fiber, is considered to be pure cellulose.

When the boll opens at the end of the growing period, the fiber dries out and the cell, no longer distended by plant juices, collapses into a shriveled, twisted, flattened tube having an average length of 2500 times its width.

Maturity of the fiber, which varies greatly within a sample, is a determining factor in the overall quality of the cotton. No commercial sample has ever been found to be 100% mature, but it is generally thought that a normal cotton with a maturity of 70% processes efficiently and produces yarns of acceptable quality. When maturity falls as low as the 50–60% range, processing difficulties result, as indicated by increased ends down in spinning, and decreased product quality.

Cotton is the most important fiber in world textile economy, representing an annual farm value of approximately 6.3 billion dollars.

Emphasis in recent years has been on the chemical modification of the cellulose in the cotton fiber to meet increasing competition from man-made textile fibers and filaments. Cotton fabrics having dimensional stability, coupled with wash-wear and durable press properties have been successfully marketed, and rot-resistant and flame-resistant materials are commercially available in 100% cotton compositions.

In the study of cotton fibers, yarns, and fabrics, the microscope has proved an invaluable tool. With this instrument it has been possible to investigate dimensions, shape, construction, markings, and other physical properties, as well as behavior toward chemical treatments which alter these properties.

The use of the microscope in the study of textile fibers dates to 1836, when a report to the Privy Council of King William IV included a microscopical study of textile fibers in Great Britain [2].

Line drawings illustrated varieties of cotton, and descriptions of their physical properties were given.

In 1907, Hanausek and Winton [3] published a book in which they showed drawings of textile fibers for identification purposes. Twenty-five years later, the microscope was in such common use for textile work that several textbooks [4,5,6,7,8] were produced giving techniques for the practice of microscopy in textile research.

Improvements in instrumentation resulting in newer and more modern designs of the light microscope have served to enhance its potentialities for morphological investigations. However, the limits of resolution imposed by the nature of the light beam led to the development, in the decade between 1930 and 1940, of a microscope utilizing an electron beam and magnetic lenses [9]. The application of this electron microscope to the study of fiber structure was first realized in 1940 [10].

At the Southern Regional Research Center the microscope is used in studies of improvements in cotton products with greatest emphasis on fabrics for wearing apparel. While a chemist can measure quantitatively the effects of a particular treatment, it is only through microscopical examinations that he can actually see the results of his efforts. Percentage "take up" of a compound can be calculated, but its location in a yarn or within fiber walls can be demonstrated only under the microscope, often after use of an appropriate dye or other marker. Among the numerous techniques employed in these studies are the use of microtomes, vacuum evaporators, various embedding and mounting media, and swelling agents.

For convenience, we have arranged the plates in the book to show characteristic features of native- and chemically-modified cottons as observed with (1) the light microscope, (2) the transmission electron microscope, and (3) the scanning electron microscope.

It is hoped that the collection will be useful in conveying a comprehensive view of cotton morphology as it relates to textile finishing.

The authors wish to express appreciation to other members of the Microscopy Unit for their cooperation in preparing this Atlas and especially to Rosalie M. Babin for her tireless efforts in obtaining the photographic enlargements.

REFERENCES

1. K. Ward, Jr., J. B. Evans, M. L. Rollins, B. Meadows, and I. V. deGruy, *Encyclopedia of Chemical Technology,* (R. E. Kirk and D. F. Othmer, eds.) Vol. 4, Interscience Encyclopedia, Inc., New York, 1949, p. 563.

2. A. Ure, *The Cotton Manufacturers of Great Britain Systematically Investigated,* Vol. 1, Charles Knight, London, 1836, p. 56.

3. T. F. Hanausek and A. L. Winton, *Microscopy of Technical Products,* Wiley, New York, 1907.

4. L. G. Laurie, *Textile Microscopy,* Ernst Benn, London, 1928.

5. J. H. Skinkle, *Elementary Textile Microscopy*, Howe Publishing, New York, 1930.

6. P. Heerman and A. Herzog, *Mikroskopische und mechanische-technische Textiluntersuchungen,* 3rd ed., Springer, Berlin, 1931.

7. J. M. Preston, *Modern Textile Microscopy,* Emmott and Co., Manchester, 1933.

8. E. R. Schwarz, *Textiles and the Microscope,* McGraw-Hill, New York, 1934.

9. S. Wischnitzer, *Introduction to Electron Microscopy,* Pergamon Press, New York, 1962.

10. M. L. Rollins, A. M. Cannizzaro, and W. R. Goynes, *Instrumental Analysis of Cotton Cellulose and Modified Cotton Cellulose* (R. T. O'Connor, ed.), Marcel Dekker, Inc., New York, 1972.

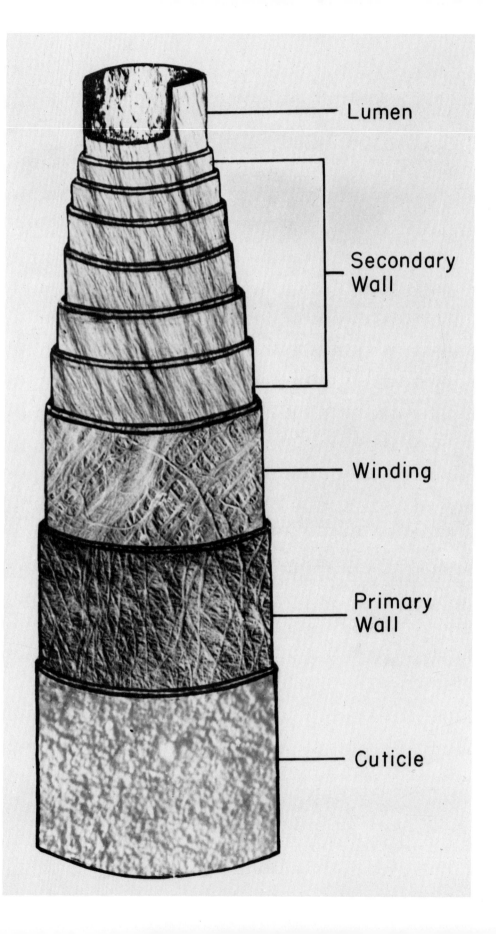

Lumen

Secondary Wall

Winding

Primary Wall

Cuticle

INTRODUCTION

THE STORY
OF COTTON

Robert T. O'Connor

HISTORY

Cotton is one of the world's most socially vital and economically important agricultural crops. Millions of individuals are dependent upon it as their main source of clothing and as an important food source. It is the sole or major source of livelihood in some stage of growing, processing, marketing, or manufacturing, for millions of people in sixty nations of the world. The lint from the cotton plant is a major source of fabric for clothing; the seed is an important supply of vegetable oil for food; the meal is a valuable supplement for feed or fodder for cattle and smaller animals, and is just now being developed as a source of vegetable protein for human diet in the form of structured meat products. All of these benefits from a single plant are probably not matched by any other agricultural commodity.

   The story of cotton predates recorded history. Long before history was written, cotton was grown and used in the manufacture of fabrics in India, Egypt, and China. There is inconclusive evidence that it existed in Middle Egypt as early as 12,000 B.C. Archeological finds of bits of cotton fiber and string at a site in the ruins of the city of Mohenjodaro in the Indus Valley in what is now Pakistan, indicate that cotton was known and used as early as 3000 B.C., and by this time natives of the Nile Valley in Egypt had mastered the art of spinning and weaving. The use of cotton in India is referred to in the ancient laws of Manu as early as 700 B.C. From as early as 1500 B.C. to the beginning of the sixteenth century, India was the world's center of the cotton industry.

Among the earliest recorded items in the history of cotton are those of Herodotus, the father of history, and of the chroniclers of the campaign of Alexander the Great (circa 300 B.C.). Herodotus wrote in 445 B.C. of the trees that grow wild in India bearing, instead of a fruit, a wool "exceeding in beauty and quality that of sheep." He described clothes made by the Indians of this "tree wool." This writer also refers to the clothing of Xerxes' army (circa 480 B.C.) as being composed of cotton fiber. Nearchus, an admiral of Alexander, refers to trees in India bearing, as it were, bunches of wool. Theophrastus (circa 350 B.C.) described in some detail the manner in which the cotton plant was cultivated in India. The use of cotton was undoubtedly known to the Greeks soon after the invasion of India by Alexander, and Aristobulus, a contemporary of Alexander, mentions the cotton plant under the name of "wool bearing tree."

Arabian travelers of the Middle Ages, writing of India, describe garments of such extraordinary perfection that nowhere else are their like to to be seen, and Arabian caravans undoubtedly first brought Indian cotton into Europe. These medieval Arabs carried cotton over the trade routes from India. Their word "Quttan" or "Kutn," meaning a plant found in conquered lands, is the origin of our "Cotton." "Muslin" originally applied to cotton woven in Mosel, also comes from the Arabic language.

Certainly, the Romans were using cotton, presumably imported from India, by 90–70 B.C. Pliny reports fairly extensive cotton cultivation and manufacture in upper Egypt in 70 A.D. The history of the use of cotton in Egypt is confused by the terms "flax" and "cotton." Herodotus states that the Egyptian priests wore linen clothes, but Pliny refers to them as also wearing cotton material. The confusion arises, apparently, from the fact that in translation the word "linen" refers not to the fiber of which the cloth was made, but to its general appearance. Thus, cloth made from either flax, or cotton alone, or from a mixture, was called "linen." Compare the present-day use of the word "lint," meaning cotton fiber. The fact that all Egyptian mummy cloths so far examined appear to consist of flax is no argument against the use of cotton at this period of history. Flax appears to have been reserved particularly for certain religious purposes, and cotton, wool, and silk may have been in common use for clothing.

By 912 A.D. cotton had established a firm foothold on the European continent, first in Spain, under the Moors, and in Sicily, under the Arabs. Later the Crusaders brought it to the attention of Northern and Western Europe.

Cotton was introduced into China and Japan from India. China was growing cotton as a decorative plant in 700 A.D. A story is told of its introduction into Japan by a shipwrecked inhabitant of India in 798 A.D. However, it was not until the seventeenth century, after its reintroduction by the Portuguese in 1592 A.D., during the reign of Tokugawa, that cultivation of cotton became general in Japan. Cotton was probably reintroduced into China at the time of the conquest of the country by the Tartars, from 1206–1300 A.D. It was not a fiber cultivated for manufacturing purposes there until after 1300 A.D.

Cotton is not indigenous only to the Eastern Hemisphere. A special variety, known as Peruvian cotton, is indigenous to Peru, and early explorers exhumed mummies wrapped in luxurious cotton cloth from the tombs of pre-Inca Peru, indicating its use as early as 200 B.C. Columbus was convinced that he had reached India when he found cotton growing in the Bahamas in 1492. In his diary he described how "The natives came swimming toward us with bales of cotton thread, which they exchanged with us." His claim that he had reached India appears to have been partially supported in the Spanish Court by the treasure he brought back to Queen Isabella—skeins of cotton thread—as cotton had been established as a product of India. Columbus' gift was "suitable for a Queen," since he had brought back Sea Island cotton, the most valuable of all species, described as the "finest growth of cotton."

Cortez found cotton used in Mexico and included many fabrics made from cotton among the presents he sent to Charles V of Spain. Thus, in 1519 the first recorded export of cotton from America to Europe was made.

By the time of Pizarro's conquest of Peru, in 1522, the inhabitants were clothed in cotton garments made from the Peruvian cotton indigenous to that country. Magellan recorded the use of cotton in Brazil in 1520.

Cotton has played a prominent role in the lives and livelihood of millions of individuals throughout the world. For perhaps 4000 years, India took and maintained her position as the leading cotton country of the world. For most of this period the crop was grown and cultivated without advantages of or advances in agronomics or genetics, and fabrics were produced with hand-operated equipment which was very crude and primitive. Despite these handicaps, Indian products, as well as raw fiber, were exported to foreign countries. The native artisans were able to weave muslins of such delicate texture that they have never been equaled in fineness. Garments made from these cloths were frequently referred to as "webs of woven wind." India remained Queen of the cotton empire until changes in world trade and shipping and the dawn of the industrial era enabled the British to lay claim to the title. By 1500 cotton was generally known throughout the world. As early as 1600, about 1500 people were employed by a thriving cotton industry in Timbuktu, in deepest Africa.

Cotton was probably introduced into England by the Crusaders. Cotton fabrics were among the wonders that attracted adventurers and traders to the East during the Dark Ages. The influence of this trade is still apparent in the English words "damask," from Damascus, "Calico," from Calicut, and "muslin," from Mosel on the Tigris in Mesopotamia, now Iraq. A series of historical events during the latter part of the sixteenth and the early seventeenth centuries was destined to plunge England into her role as the dominant cotton-manufacturing country of the world. The first of these was the defeat of the mighty Spanish Armada in 1588. With supremacy of the seas assured, British sailors were soon blazing a path to the East over which the white gold was soon to travel. The second factor was the inventiveness of her people which was to usher in, in England, the prelude to the Industrial Revolution and, with it, England's supremacy as an industrial nation. However, the impact of these two events had to await still a third, the Manchester Act of 1736, delaying England's rise to dominancy as the manufacturer of cotton goods until about the middle of the eighteenth century.

The early eighteenth century was fraught with dissension and resistance to changes to the English economy. The wool industry was inflamed in protest and wool manufacturers succeeded in forcing a fiber monopoly and in influencing Parliament to pass, in 1700, a law forbidding the sale of cotton goods in England. In 1712 a companion act was passed prohibiting the wearing of all printed goods, regardless of fabric. A fine of 5 pounds for wearing cotton, and 20 pounds for selling it, was imposed upon the people. However, popular resentment at being told what to wear and how to dress by Parliament, and the continued growth of a cotton industry, despite prohibitions, in the cities of Manchester, Bolton and Lancashire, with the resulting prosperity from cotton manufacture, forced the passage of the Manchester Act in 1736. This act lifted the 24-year-old ban and permitted the wearing of cotton and linen mixed calico, marking the beginning of England's world dominance in cotton manufacturing that was to go unchallenged for almost 200 years. However, progress was not made without problems. English workers, dependent upon hand work for livelihood, immediately recognized the implications of the first products of inventive genius.

In 1730, John Kay, paradoxically son of a woolen manufacturer, invented the flying shuttle and increased production from cotton looms fourfold. This invention enabled English weavers to produce broad woven fabrics matching those of India and ignited the spark that was to set off the industrial revolution and, along the way, make cotton the king of the textile world. Records reveal that the flying shuttle invented by John Kay was ruthlessly stolen by the woolen manufacturers, and that he fell a victim to the mad fear which pervaded textile England. His invention was destroyed by mob violence and his life saved only through a hasty and secret retreat. He sought refuge in France, where he died in poverty in 1764. But the age of invention, the prelude to the Industrial Revolution, had begun. From the middle of the eighteenth century, the history of cotton in England is largely a history of mechanical invention and the application of power to machinery. The machine age, which was to establish the new principle of division of labor and specialization in various phases of production, and in this way destroy individualized craft expression and the age of the guilds, had begun.

James Hargreaves invented the first practical spinning frame in 1770, and a year later Richard Arkwright perfected the roller spinning method using horsepower in the world's first cotton mill in Nottingham. By 1790, over 150 mills were operating on this system and Arkwright had become a millionaire to be remembered as "the father of the textile industry." In 1779, Samuel Compton invented the spinning mule, giving England complete control over the difficult art of spinning cotton. This invention is still the basis of our modern spinning frame and, as a sidelight to history, although thousands of people became wealthy through this mechanical contribution, its inventor was left in poverty.

The last of the great forefathers of the English textile industry was Edmund Cartwright who, in 1785, gave the world its first automatic power loom. His first harnessing of the engine and other improvements on earlier methods of spinning revolutionized the textile industry. These outstanding contributions to create a textile industry were followed in succeeding years by improvements increasing speed and production. Hand labor passed into history—the machine age was born. England was the leading producer of manufactured cotton products and had ceased to be an agricultural nation.

Cotton was first planted by the American colonists in Virginia as early as 1619, and English wool merchants were complaining about cotton imports into England from America as early as 1621. Records reveal that by 1700 cotton was being used for the clothing of about one-fifth of the inhabitants of North Carolina. In 1748 Georgia exported the first consignment of cotton to England; in 1753 South Carolina was sending small amounts to London; in 1784, when fourteen bales of cotton arrived in Liverpool from America, eight were seized on the grounds that so much cotton could not have been produced in America. The invention of the cotton gin by Eli Whitney in 1793, which enabled a single worker seeding two pounds of cotton per day by hand to seed several hundred, provided the spark for the agronomic cotton-growing South; it was comparable to the spinning and weaving mechanical devices which had sparked cotton manufacturing in England. Cotton planting rapidly expanded through the Southern states during the early part of the nineteenth century: from Virginia, the Carolinas, and Georgia, first into Alabalma, Mississippi, and Louisiana, and later into Texas, Oklahoma, and, finally, to Arizona and California.

The fact that America was blessed with ideal climate and soil conditions for the production of cotton, as she had been similarly blessed for the growth of other agricultural crops throughout her broad lands, is evidenced by the fact that cotton has been and still is raised as an important commercial crop in 18 of the 50 states of the nation. The country is usually divided into four cotton-growing regions (1) Delta, including the states of Missouri, Arkansas, Tennessee, Mississippi, Louisiana, Illinois, and Kentucky; (2) Southeast, including Virginia, North Carolina, South Carolina,

Georgia, Florida, and Alabama; (3) Southwest, encompassing the states of Texas, Oklahoma, and Kansas; and (4) West, comprising the states of California, Arizona, New Mexico, and Nevada.

By the time of Washington's inauguration, the Southern states were raising 3000 bales of cotton per year. By 1825, that had increased to 533,000 bales; by 1850, to 2,136,000 bales; and by 1875, to 18,123,000 bales. England was, by now, importing from America the major portion of raw cotton to feed her cotton mills. The South had become the world's leader in the production of cotton. By 1926, a record-breaking 17,978,000 bales were produced. Since 1929, the effects, first of the depression and, later, of the Commodity Credit Corporation curtailment of acreage, resulted in a reduced crop, but by 1931, in the midst of depression, production was 17,097,000 bales.

Since 1931, for a period of about three and one-half decades, or until the middle of the 1960s, annual production has been maintained at around 14,000,000–15,000,000 bales per year, the increased curtailment of acreage being largely offset by increased production per acre. During the 1930s, the average of 13,146,000 bales was produced on about 32,252,000 acres. During the first half of the 1960s, about the same amount was being raised on only 15,700,000 acres. The increase per acre has been attributable to several factors, mainly to the shift to more fertile lands. With cotton acreages limited by allotments on farms there has naturally been a tendency to concentrate plantings on the best lands. In general, the increased yield per acre can be attributed to: (1) land selection and improvement; (2) soil preparation; (3) timely planting; (4) improved planting seed; (5) fertilization; (6) weed control; (7) insect control; (8) irrigation; (9) defoliation; and (10) timely harvesting. In average yields per acre, this increase has been from 170 pounds in 1930 to a level of 500 pounds per acre in 1960.

Production fell off about the middle of the 1960s, reaching a low of 7,458,000 bales in 1967 . Since the low production year, total yield appears to have leveled off again at around 10,000,000–11,000,000. In 1971, 10,473,000 bales were produced in the United States, representing about 18% of the world's 57,000,000 bales. Mill consumption for 1970 was 8,193,000 bales, and exports accounted for 3,897,000 bales.

The rise of the South as the leading producer of cotton is, however, only part of America's contribution to the cotton story. As early as the pre-Revolutionary War period, extensive attempts were made to establish a textile industry in the South. In retrospect, these attempts were premature. The South had neither the physical nor the manpower resources, nor the desire to become a manufacturing center. The rise of the English textile industry offered an unlimited market for Southern cotton and, coupled with the fact that this region was blessed with ideal conditions to produce raw cotton, made it much more advantageous for the South to supply raw cotton to others than to compete in finished production. With such abundant wealth so easily obtainable, all thoughts of industrialization were soon minimized as the plantation economy arose throughout the South as the shortest and surest road to prosperity. Travel along this road shortly made the South the major source of raw cotton throughout the world.

If the South, in the latter part of the eighteenth century, through a combination of circumstances, inclinations, and lack of resources, had no interest in becoming a manufacturing center, the North, particularly New England, did. The ingredients which the South lacked, natural power, skilled labor, and mechanical ability, all favored industrial expansion in the North. There does not appear to be a reliable record of the first successful cotton mill in the North. Among the early mills was an establishment in Philadelphia, in 1775, which operated for only two years before it was destroyed by the British. A mule-powered mill was erected in Beverly, Massachusetts, and records reveal that it was visited by President Washington in 1789. It remained in operation until 1815. A mill established in Pawtucket, Rhode Island, in 1793, is frequently referred to as representing the birth of the modern textile industry in America. It was established by Samuel Slater, who has become known as the "Father of American Manufacturing." Slater was an expert machinist in the first English mill completely equipped with Arkwright's water-powered system of cotton processing. Although English law at the time did not permit machinists to leave England, Slater posed as a barber, reached this country, and reproduced Arkwright's spinning machinery from memory. Slater rapidly took the lead in the expansion of cotton spinning mills in New England and lived to see the industry grow from his single mill to over 100 mills. The invention of the cotton gin by Eli Whitney at this time had increased the production of cotton in the South. In less than 20 years, from 1793–1811, cotton exports had increased from 487,000 pounds to over 62,000,000 pounds per year, and the cash value of the crop had increased, in the single decade from

1794–1804, from $150,000 to over $8,000,000. The invention in 1828 of the ring spinning frame by John Thorpe gave the infant cotton manufacturing industry in the North the advantage it needed, coupled with the unlimited supply of cotton from the South, to challenge world competition. This invention made spinning faster, simpler, and less expensive. Coupled with Yankee ingenuity, cotton mill operations were further perfected, and with the opening up of the seas by Yankee sailors, the combination of the cotton-producing, agronomic South, and the cotton-manufacturing, industrial North shortly took preeminence of the cotton textile industry away from England, just as she had wrested it from India about a century earlier.

Now that cotton could be ginned in volume, and now that it was being grown in abundance in the South, spinning and weaving grew into a tremendous industry in the North. A Boston merchant accelerated this growth by matching, in weaving processes, what Slater had accomplished for spinning. As Slater had done earlier, Lowell reproduced the machinery from memory, completing the spinning process from bale to woven fabric under one roof in his Waltham, Massachusetts Mill. By 1816, 500,000 spindles in cotton mills were employing more than 100,000 workers and represented an invested capital of $50,000,000.

The combination of the agronomic cotton-producing South and the industrial cotton-manufacturing North was destined to change. Although the South had turned, after the first efforts at industrialization in the early days of the eighteenth century following the Revolutionary War, to a plantation economy, individual leaders rose to become crusaders for industrial expansion. William Gregg, noted as the South's greatest pioneer in large-scale textile manufacturing, erected a mill in Graniteville, South Carolina, in 1845, which is still operating. Students have argued that Gregg's social programs augmented the drive toward industrialization of the South. Certainly he was a crusader. He prohibited child labor, made school attendance compulsory, furnished free text books to children of his employees, refused to use slave labor, and incorporated into his mill operations theories and practices which won him fame as an industrial leader in the South. His success served to bring other Southern businessmen to the realization that industry could and probably should flourish there. The Cotton Exposition of 1881 in Atlanta served to further reinforce this thinking. However, it was the Civil War and its aftermath which ushered in the industrial expansion of the South. Historians can debate the exact causes: the breakdown of the plantation system, the change from slave to free labor, the complete collapse of commerce and industry, the frustration and futility of the reconstruction period, the general era of change, whether welcome or not, all probably played a part. But whatever the cause, by 1895 the South had entered into an industrialization which was to expand probably beyond even the dreams of the men of vision of the New South.

The rapid expansion of Southern industrialization, as measured by mill consumption of raw cotton, is illustrated by Table 1, which compares, over a period of 50 years, the proportion of U.S. total mill consumption by 10 Southern states with that of 13 Northern states, together having accounted for about 98% of the country's mill consumption.

In 1905 the industrial North still led the South, but two decades later, in 1925, the South had increased the ratio over the North to about 2–1. Another two decades later, in 1945, the South's lead had increased to 8–1, and in 1960 the 10 Southern states accounted for 94.8% of the nation's mill consumption. During the early 1960s, the country's mill consumption averaged around 9,000,000 bales annually, with 4 Southern states (North Carolina, South Carolina, Georgia, and Alabama) accounting for about 90% of the total. The New South reigned supreme as King Cotton's Queen, successfully defending her title against all claimants until the present day.

Cotton production is today America's greatest agricultural

Table 1

MILL CONSUMPTION
(COMPARISON OF SOUTHERN AND NORTHERN STATES
FOR SELECTED SERIES OF YEARS)

Year	U.S. total consumption (1000 bales)	Ten Southern states consumption (1000 bales)	Percent	Thirteen Northern states consumption (1000 bales)	Percent
1905	4909	2346	47.8[a]	2507	51.1
1925	6456	4449	68.9	1964	30.4
1945	9163	7983	87.1	1044	11.4
1960	8252	7853	94.8	2300	2.8

[a]Percentages do not add to 100, as about 2% mill consumption in minor mill consumption States has not been included.

industry. Currently more than $1,470,000,000 annual income, or 3% of the total value of all United States farm crops is derived from cotton and cottonseed. About 11% of the Nation's population, or some 18,200,000 people, make all or part of their living by growing, processing, manufacturing, and selling cotton and its products. More than $13,000,000,000 are invested in the raw cotton industry. Cotton is grown in 18 of the 50 states and a network of over 7080 textile establishments is located in 27 states from Maine to Texas. In the post World War II period, United States cotton textile and fiber exports have averaged about 1,923,807,000 pounds, making the Nation one of the world's largest exporters.

But more and more today, cotton is not only of importance to America; it is one of the world's most vital and necessary crops. Its production and sale constitute one of the major factors in world prosperity and economic stability. Millions of individuals are dependent on cotton for livelihood and economic dependence. Cotton is grown in 60 countries throughout the world in increasing quantities, and mill consumption is likewise increasing.

As shown in Tables 2 and 3, world production of cotton has increased considerably (about 56%) from the 1934–1938 period to the present year: from about 32 million to 57 million bales. United States production has declined slightly in the number of bales (from about 13 million to 10 million) but has declined drastically in percent (a 24% drop in the percent of the world's production, from 42% in 1934–1938 to 18% in 1971–1972). In the period 1934–1938, the United States produced 42% of the world's cotton, an amount almost equal to all countries of the world, excepting Russia, and almost four times that of any one country. By 1971–1972 it was producing only 18% of the world's supply, had yielded first place in world production to the U.S.S.R., and was actually producing only twice that of China, India, and Brazil, the three other leading cotton producing countries.

World consumption of cotton has shown an even more drastic increase during this period, an increase of from about 30 million bales to 55 million bales, about 83%, from 1934–1938 to 1970–1971. The United States shows a relatively modest increase from 6 million bales to 8 million bales, but a drop in the world's relative consumption from over 22% to about 15%. The United States, during the period 1934–1938, consumed twice as much cotton as Russia, China, India, Japan, and the United Kingdom, the five other leading countries of the world. By 1970–1971, consumption in this country

Table 2

WORLD COTTON PRODUCTION DURING SPECIFIC PERIODS
(1000 BALES)

Country	1934–1938	1945–1949	1955–1959	1960–1964	1965–1968	1970–1971	1971–1972
World	31,689	25,687	43,888	48,276	51,039	52,486	57,076
U.S.A.	13,149 (42%)	12,104 (47%)	13,013 (30%)	14,795 (31%)	10,589 (21%)	10,269 (20%)	10,250 (18%)
U.S.S.R.	3430 (11%)	2328 (9%)	6750 (15%)	7370 (15%)	9140 (18%)	10,800 (21%)	11,000 (19%)
China	2855	1939	7360	4940	6740	7800	7600
India	5348	2304	3991	4741	4870	4400	5800
Brazil	1956	1352	1490	2235	2730	2300	3100
Pakistan		1024	1376	1656	2262	2570	3350
Egypt	1893	1456	1807	2037	2198	2346	2351
Turkey	249	268	734	1091	1777	1845	2400
Mexico	302	577	2032	2206	2215	1440	1715
Sudan	439	371	500	675	1156	1130	1100
Total eight countries	13,074 (41%)	9291 (36%)	19,290 (44%)	19,581 (41%)	23,948 (47%)	23,831 (45%)	27,416 (48%)
Fifty other countries	6%	8%	11%	13%	14%	14%	15%

Table 3

WORLD MILL CONSUMPTION
DURING SPECIFIC PERIODS
(1000 BALES)

Country	1934–1938	1950–1951	1960–1961	1963–1964	1970–1971	1970–1971 Percent
World	29,609	35,080	47,986	47,544	54,964	—
U.S.S.R.	3058	3950	6200	6600	8500	15% } 30
U.S.A.	6434	10,509	8279	8609	8068	15%
China	3600	3250	6800	5500	8200	14% } 29
India	3096	3150	4620	5250	5200	9%
Japan	3315	1599	3441	3164	3219	6%
Pakistan		150	1125	1240	2025	— } 10
Brazil	512	840	1250	1250	1380	—
West Germany	1077	1050	1500	1312	1078	—
France	1181	1255	1397	1307	1095	—
Egypt	73	281	550	610	935	—
Italy	684	987	1040	1049	925	—
Turkey	97	230	589	560	850	— } 9
Hong Kong		127	480	575	801	—
United Kingdom	2741	2135	1232	1065	741	—
Mexico	227	335	500	560	675	—
Other						22

was somewhat greater than that of Russia and China. The U.S. Department of Agriculture estimate of the cotton crop for 1972 is 13.6 million bales, which, if achieved, would mean the largest U.S. production since 1965 when 14.3 million bales were produced.

The U.S.S.R. and U.S.A. accounted for about 30% of the world's consumption, and China, India, and Japan, together, accounted for another 30%. These tables document the importance of cotton production and consumption to the economic welfare of many countries of the world.

Cotton is a member of the botanical genus *Gossypium,* which includes the cultivated cottons and is a member of the subtribe *Hibisceae* in the natural order of *Malvaceae.* Truly wild forms of cotton also belong to this genus. Several species exist. Sydney Cross Harland's botanical classification includes 18 species, both cultivated and wild, and does not provide for some species admitted by other workers. These exclusions were made either because particular cases were so close to accepted species as to merit only sub species, or varietal rank, or because insufficient knowledge was available to establish valid individual species. Harland describes *Gossypium* as follows:

> Habit ranges from that of a herbaceous plant to subarboraceous shrub or small tree. Main stem round and characterized by a lower zone with monopodial branches and an upper zone with symbodial branches. Flowers are cream, yellow, red, or purple, and are borne on sympodial branches. Bracteoles three, large or small, cordate, toothed, or entire. Calyx truncate or five-toothed. Staminal column bearing indefinite filaments, below naked or with anthers to the apex. Ovary 2–5 locular, seeds per loculous indefinite. Styla glandular, club-shaped or clavate shortly into as many lobes as loculi in the ovary. Capsule with loculicidal dehiscense. Seeds subglobose or angular, covered with one or two layers of unicellar convoluted hairs: albumen thin, membranous or absent; cotyledons strongly folded. Glands nigro-punctate distributed over the whole plant. Leaves entire or 3–9 lobed.

Cotton is indigenous to both the Eastern and Western Hemispheres. Investigations appear to indicate that there were probably four general centers of origin of the cotton plant, two in the Old World (Indochina and tropical Africa), and two in the New World (Mexico or Central America, and the foothills of the Andes Mountains in South America). The cultivated cottons of today can, it appears, be traced back to cottons grown in ancient times in one or another of these four world centers.

More recent studies of cotton from different parts of the world indicate that the independent origin of American (New World) and Asiatic and African (Old World) cottons has remained distinct. They are so incompatible that cross breeding is rare and persistent fertile hybrids are unknown. The New World, American, cottons have 26 chromosomes; Asiatic and African, Old World, cottons have only 13.

Four species of Gossypium account for practically the world's supply of cultivated cotton. These species were independently developed and cultivated by different tropical peoples in both the Old and New Worlds.

WESTERN HEMISPHERE—NEW WORLD COTTONS

G. hirsutum, of Central American origin, is a shrubby plant which reaches a maximum of 6 ft. It is the species containing the Upland varieties of cotton. These are the most important varieties of cotton in world commerce, constituting 87% of the world's production. Harland describes these most important varieties of cotton:

> Habit sympodial, the first fruiting being produced at nodes 6–10. Boll pale green and with glands buried beneath surface, more often round than long; sometimes very large. Anthers early bursting. Filaments long at top and middle of column, shorter at base. Pollen medium yellow or medium cream, spotted in some types, but usually devoid of spot.

Since the Old World and the New World cottons are not known to hybridize under conditions of cultivation, all of the Upland varieties now grown in this country are quite definitely of the American type and probably all came from Mexico and Central America. The vast differences in climate and soil that obtain over the Cotton Belt undoubtedly brought about a kind of natural selection which eliminated many of the varieties that were tried, while others became adapted to the several conditions under which they were grown and selected over a period of years.

As a result of such adaptation and selective breeding, there were many varieties of Upland cotton in existence when cotton first became important as a commercial crop in the middle of the eighteenth century in the Southern portions of the United States. In the eastern edge of the Cotton Belt, varieties were found that were smaller bolled, prolific cottons, characterized by green seed and softer fiber, and probably originating in the more humid eastern portion of Central America and Mexico. On the western edge of the Cotton Belt, cottons from the interior or drier section of Mexico were developed. These types were characterized by large white seed, large bolls, and hard fiber. Through selection and crossbreeding a large number of varieties of Upland cotton were developed, the selection, at first, being made for production, that is, yield per acre.

The spread of the cotton boll weevil over the Cotton Belt at the beginning of the present century, however, caused considerable change in the types of cottons selected for growth in the different sections. The earlier selected varieties in much of the Cotton Belt were the later maturing varieties of better staple than earlier short-staple types and were much more productive. When the boll weevil struck these areas, however, it was no longer possible to grow these late-maturing varieties. Varieties that had become famous for high quality, but were late maturing, were discarded, and early maturing, shorter-staple cottons were substituted. These were usually inferior in quality but had developed early maturity, that is, they fruited rapidly instead of over a long season. They also were varieties which afforded a satisfactory production in terms of yield per acre. There became only two criteria for selection and breeding: first, that the variety must be early maturing to be boll weevil resistant and, second, that it must be productive, that is, have a high yield per acre to meet economic requirements. In this way many excellent varieties of long-staple Upland cotton and practically all of the better types of medium staple were lost within a comparatively short time, to be replaced by the early, rapid-fruiting varieties.

The realization that these early maturing cottons were of very inferior quality and that their production was resulting in the loss of markets which had been using the better cotton of pre-boll weevil days for many years, led to a more scientifically planned breeding program by the U.S. Department of Agriculture in cooperation with State Experiment Stations and several individual plant breeders. The objectives of this program were to develop varieties of Upland cottons with the economic advantages of high productivity in yield per acre and high resistance to attack by insects and by several of the cotton plant diseases, such as cotton wilt, while maintaining improved staple length, fineness, and overall quality-properties.

These programs produced a very large number of varieties of Upland cotton. As early as 1910, the U.S. Department of Agriculture published a bulletin listing over 600 varieties. By 1962, however, 7 varieties were accounting for 70% of the cotton crop and 40 varieties for more than 98%. The program produced a balance between the economically necessary and technologically desirable properties of Upland cotton. Production was increased, as mentioned earlier, from an average of 14,500,000 bales on about 42,000,000 acres in the 1930s to production of about the same number of bales during the 1960s on only 15,700,000 acres. Varieties resistant to insects, particularly the boll weevil, and to cotton diseases, such as the cotton wilt, were developed. Along with these developments the staple length of Upland cottons was steadily increased and overall quality, in terms of such factors as harvesting, ease of ginning, preparation for spinning, and performance of the fiber when put to the final test in manufacturing processes was improved. Staple length, in particular, was further improved by crosshybridization of the Barbadense, particularly the Sea Island and Egyptian varieties, with the American Upland cottons, enabling them to compete in expanded markets once limited to the finer Sea Island and Egyptian varieties of the *G. barbadense.* The success of the program was probably the major factor in the increased production of American Upland cottons to the stage where they account for 87% of the world's production, as previously mentioned.

G. barbadense, believed to have originated in Peru, grows from 6–15 ft. Although including some of the Egyptian varieties of cotton, which gained popularity in filling demands for finer count, long-staple fiber, *Barbadense* cotton is not native to Egypt. It is a New World cotton, having been introduced into Egypt from America early in the eighteenth century and cultivated as the Egyptian variety along the Nile. The growing of Egyptian *Barbadense* spread to many parts of the world including Sudan, Peru, the U.S.S.R., and the United States, where it became known as American-Egyptian.

Another variety of *G. barbadense* is the famous Sea Island cotton which was of considerable importance in the early American development of cotton cultivation. It became of particular value as a source of supply for the "top quality" fraction of the trade whose needs were not met by the higher yielding, shorter staple of the American Upland varieties. The Sea Island variety was introduced from the Bahama Islands; it was developed on islands off the coasts of South Carolina and Georgia, and on narrow strips of the Mainland near the coast. It became recognized as the "finest quality of cotton known," and survived for about 100 years with this reputation, until gradually certain varieties of Egyptian cotton gained ascendancy in supplying the demand for fine cotton. The Pima cottons, now cultivated in the Southwest portions of the United States and in Northern Peru, and other varieties of *G. barbadense* descended from the Egyptian long staples. Pima S-2, one of the most popular of the current plants of these varieties, includes, other germplasma in its ancestry.

G. barbadense, which grows from black seed, constitutes about 8% of the world's cotton crop. Thus, New World cotton accounts for a total of 95% of the world's production.

EASTERN HEMISPHERE—OLD WORLD COTTONS

G. arboreum includes the tree cottons and some of the native cottons of India and Pakistan. It grows as tall as 15–20 ft. Its seeds are covered with a greenish fur.

G. herbaceum averages 4–6 ft. and is grown from a seed encased in a gray down.

Asiatic cottons belonging to *G. arboreum* and *G. herbaceum* once played a major role in the history of cotton. They are now disappearing because of their low yields and short fiber, surviving mainly in India and Pakistan. These species produce approximately 5% of the world's crop, most of which is consumed locally. A small quantity of a very coarse fiber from *G. arboreum,* known as Desi cotton, is exported for special needs, such as mixing with wool.

In the study of cotton fibers, yarns, and fabrics, the microscope has proved to be an invaluable tool over a long period of years. It has been performing, for cotton, the type of analyses referred to by Dr. W. Wayne Meinke, Head of the Analytical Chemistry Division of the National Bureau of Standards. This is particularly fine for surface effects and for the examination of the inner structure of the fiber.

Emphasis in recent years has been on the chemical modification of the cellulose in the cotton fiber to meet increasing competition from man-made textile fibers and filaments. Cotton fabrics having dimensional stability, coupled with wash-wear and durable-press properties, have been successfully marketed, and rot-resistant and flame-resistant materials are commercially available in 100% cotton compositions.

In the development of such modifications of cotton fibers, yarns, and fabrics, the microscope has proved an invaluable tool. Preliminary analyses to determine the success of specific processes in initial experiments and quality control in subsequent production have contributed to the success of this revolutionary advance. At the Southern Regional Research Center of the Agricultural Research Service of the United States Department of Agriculture, the microscope is used in studies of improvements in cotton products, with the greatest emphasis on fabrics for wearing apparel. With the microscope, it has been possible to investigate dimensions, shape, markings, and other physical properties of the cotton fiber, as well as behavior toward chemical treatments which alter these properties. While a chemist can measure, quantitatively, the effects of a particular treatment, it is only through microscopical examination that the results of his efforts can be seen. Percentage "take up" of a compound can be calculated, but its location in a yarn or within fiber walls can be demonstrated only under the microscope. Numerous techniques employed in these studies include the use of microtomes, vacuum evaporators, various embedding and mounting media, swelling agents, and appropriate dyes.

1

LIGHT
MICROSCOPY

The origin of the light microscope dates back to the sixteenth century, but Anthony van Leeuwenhoek, in Holland, in the seventeenth and eighteenth centuries, was the first to design and use the simple microscope for serious studies. Robert Hooke, of England, later used the double telescope eyepiece designed by Christian Huyghens, a Dutch mathematician and physicist, and was credited by the Royal Microscopical Society as being the inventor of the true compound microscope.

Today the light microscope is used extensively in the study of textile fibers and its effectiveness is enhanced by the development of such techniques as staining, swelling, and sectioning.

FIBERS

Examination of fibers, the basic units of textiles, has been and remains the largest part of the work of the textile microscopist. To understand changes which cotton undergoes during various chemical processes it is necessary first to understand the morphology of the fiber as it appears under the microscope. Differ-

21

34014

ences between mature and immature fibers are demonstrated by the transparency of fiber walls in immature fibers as compared with the solid appearance of mature fibers.

Stains and swelling agents often make it possible to show types of damage to fibers from mildew, heat, bacteria, acid, abrasion, etc. Cracks in the primary wall can be shown by concentration of a particular stain in selected areas and/or breaks in these areas when certain swelling agents are applied to the fiber.

YARNS

When fibers are spun into yarns, the number indicates the yardage required to make one pound of a particular size yarn. For example, a 60s yarn is one whose number, when multiplied by 840 gives the yardage of the yarn in one pound (840 x 60 = 50,400 yards per pound). A cross section of such a yarn examined under the microscope reveals size, shape, and fiber relationship within the yarn, as well as percentage amounts of mature and immature fibers comprising the yarn.

In studies of chemically treated yarns selective stains are often used to show location of added material. When polymers have been deposited in fibers or between fibers, a dark stain which has an affinity for the polymer but not for the cotton is applied and a yarn cross section is made to demonstrate the location of the polymer. Differences in methods of applying the same polymer with varying results are also compared microscopically. This type of cooperative research is of value to the chemist who can assay his results and alter compounds or methods of application to improve the ultimate product.

FABRICS

During the last half century the solution of mill processing problems and the explanation of fundamental principles of textile processing have been aided considerably by the use of the microscope.

It is possible to show through microscopical techniques some of the physical characteristics responsible for a cotton having good or poor spinning qualities.

Behavior of fabrics under conditions of use can be studied and often explained by microscopic examination of weave, construction, and interyarn relationships. Effects of new developments in textile machinery can sometimes be demonstrated by low power photomacrographs of fabric surfaces. Such examinations reveal gross details of construction and fabric geometry, whereas observation of thin sections cut through warp or filling of a woven fabric establishes fine constructional details. Median sections cut through a flat fabric add still another dimension to the study of internal structure of woven fabrics.

PLATES

PLATE 1.

Longitudinal view of immature cotton fibers with transparent, ribbon-like appearance, resulting when thin-walled tubes collapse.

These fibers were mounted on a slide in paraffin oil because the refractive indices of the fiber and liquid are such that they complement each other and enhance the visible characteristics of the fiber.

Immature fibers of the type seen in this photograph are of great concern to the mill processor since they are responsible for the formation of neps or badly tangled knots of fibers which cling together throughout processing and later cause imperfections in yarns and spotty dyeing. A high percentage of these fibers in the raw stock results in an excessive number of neps, increased ends down in spinning, and, ultimately, a fabric which must be rejected by the finisher. However, if the percentage of immature fibers is low, they do not cause trouble since they are usually blended with mature fibers during carding and combing in the mill.

Samples of raw cotton with a maturity count of 80–90% are considered of a high grade; normal cottons have a maturity of 70%; cottons of maturity below 60% result in processing difficulties and produce yarns of unacceptable quality.

Magnification x625

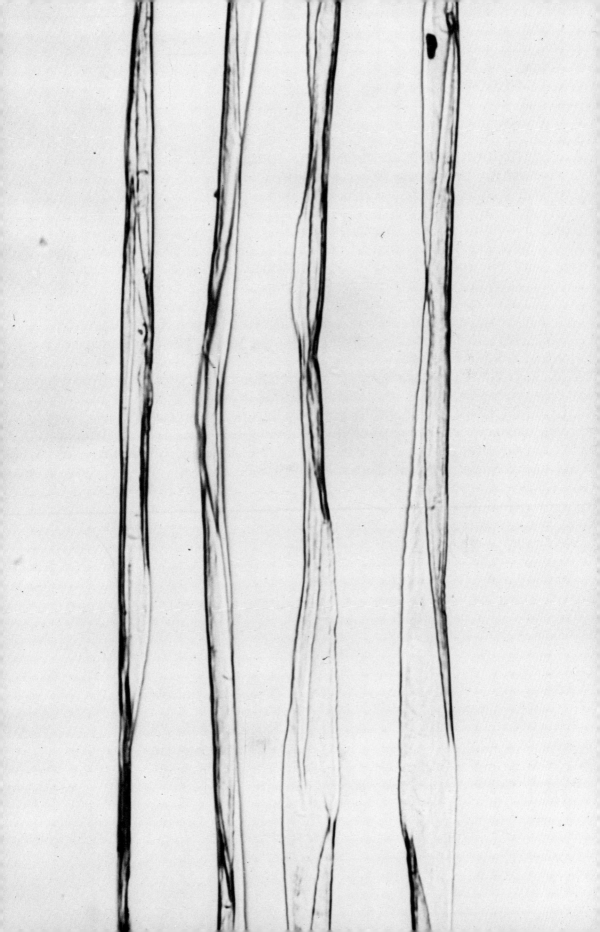

PLATE 2.

Mature cotton fibers having shapes different from those of the immature fibers in Plate 1.

Mature fibers have a thick, full-bodied appearance because of the heavy deposition of cellulose in the secondary wall. During maturation, cell walls vary considerably in degree of thickening, causing extreme variability in the cross-sectional sizes and shapes often observed in commercial lint.

Except for the base and tip, the mature fiber is essentially the same throughout its length, although it twists to varying degrees giving rise to convolutions.

There is no known single explanation for the cause of convolutions, but it has been suggested that the underlying spirality of the fibrillar elements of the secondary wall brings about this formation. The presence of these convolutions greatly affects the tensile strength and birefringence of the fiber, as well as many of the other physical properties.

Magnification x625

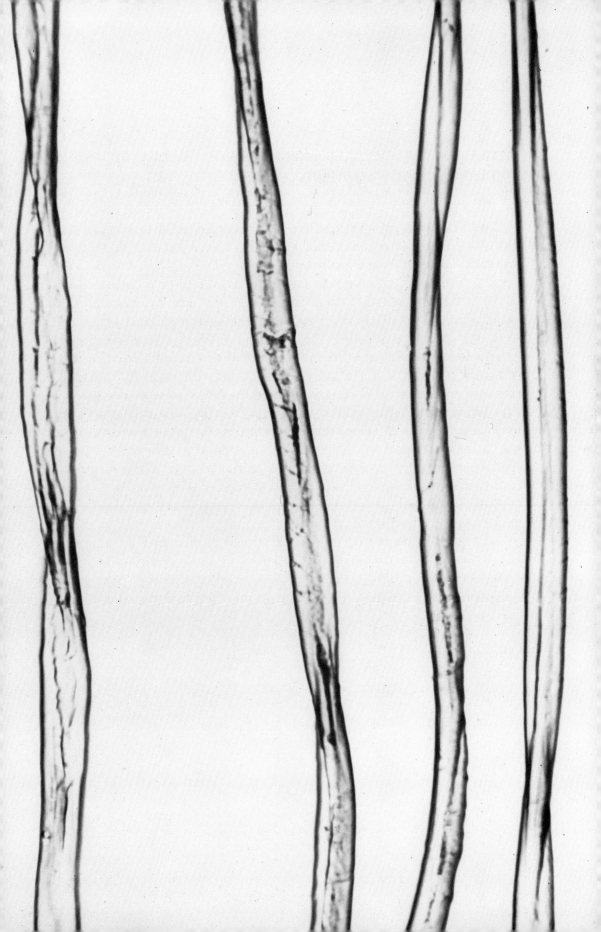

PLATE 3.

Cross section of typical Upland cotton fibers prepared in the Hardy hand microtome.

The fiber bundle is embedded in nitrocellulose lacquer and sections are sliced approximately 7 μ thick. Note varying degrees of maturity as seen in cell wall thicknesses.

The most striking characteristic of cross sections of cotton fibers is the variability of every dimensional feature. Their cross-sectional shapes range from circular, to elliptical, to linear, and often are referred to as resembling kidney beans. The general geometric shape for most mature fibers is elliptical to circular; for the less-thickened fibers it is flat and rectangular with rounded corners. Immature fibers are often U-shaped in cross section because of the tendency of thin-walled fibers to fold or curl on themselves.

The lumen of the fiber varies in dimensions over a wide range, as is demonstrated by this cross section. Mature fibers may be so fully developed as to completely close the lumen, while others of intermediate maturity have ellipitcal to circular openings in the center. Extremely immature fibers are completely collapsed and the lumen wall adheres to itself.

Although there is great variability in fiber size and shape, occasionally samples of unusual uniformity in cross-sectional features are encountered. These cottons have a high maturity count, somewhere between 70–80% level, but are not often found in day-to-day processing.

Magnification x1125

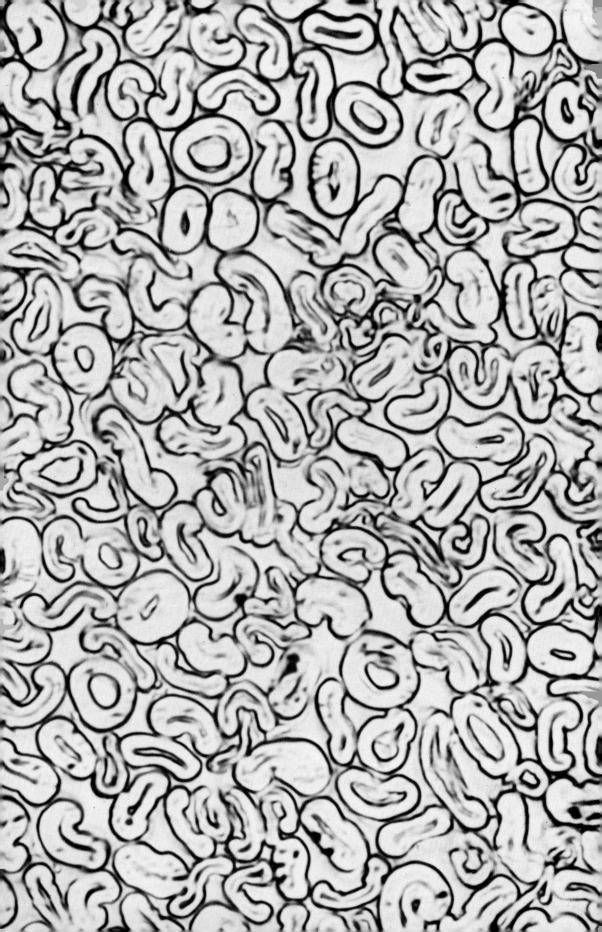

PLATE 4.

Cross section of mercerized cotton fibers showing rounded shapes and tightly closed lumens.

Mercerization is a widely practiced process whereby cotton is treated by immersion in 20% sodium hydroxide solution at room temperature. This modification of cotton is employed commercially to improve luster and strength of yarns and to enhance the dyeability of cotton materials.

The commercial mercerization process is usually very rapid with the result that it is incomplete and the morphological features of raw cotton are not greatly changed. However, when cotton is fully mercerized it swells and becomes almost circular in cross section, and in the fully mercerized mature fiber, the lumen is virtually closed. On washing and drying, the general features of the mercerized fiber remain the same except for a small change in cross-sectional area.

Variations in mercerizing processes include such factors as dwell time, concentration of solution, and degrees of tension. Most manufacturers agree on the need for rigid tension control since it has such a tremendous effect on luster, strength, and elongation—all important textile properties.

Magnification x1000

PLATE 5.

Longitudinal view of cotton fibers showing wedge-shaped and lateral cracks in fiber walls from overexposure to excessive heat (162°C for 24 h).

An understanding of the process by which cotton is degraded by heating in the presence of moisture and air is important in finishing processes, in laundering, and in connection with certain end-uses where elevated temperatures are involved. Conveyer belts and ironing-board covers are typical examples of this.

Overheating produces permanent effects on such properties of cotton as elongation-at-break, degree of polymerization, and breaking strength. It has been found that heating in the presence of moisture and oxygen increases the degradation of cellulose, but when both are removed from the system, this does not occur.

Detection of heat damage by microscopic techniques involves the use of agents which swell the secondary wall cellulose; the pressure causes breaks in the fiber at weak points along its length which become pronounced when the fiber is immersed in a caustic solution. The wedge-shaped cracks in this photomicrograph demonstrate this type of damage.

Magnification x540

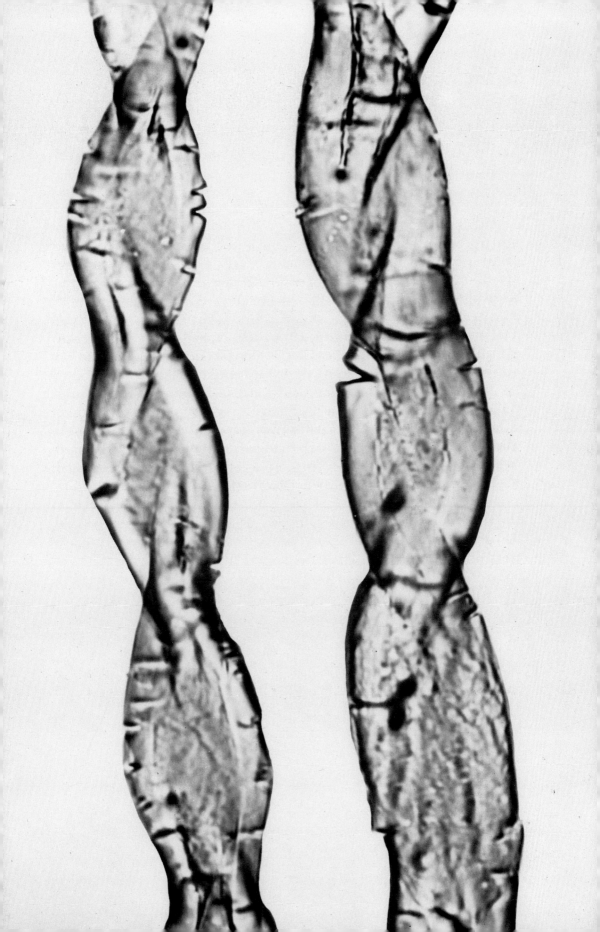

PLATE 6.

Longitudinal view of acetylated cotton fibers showing slight damage to primary wall.

Several properties are induced or augmented by partial acetylation of cotton, such as resistance to heat degradation and reduction of rate of bacterial decay. While many of the phenomena associated with the chemical modification of cotton take place at the submicroscopic level, the effects of the modification can become apparent by examination under the microscope. Visualization of physical changes is perhaps the most important contribution microscopy can make in evaluating the total effect of chemical modification on the physical properties of the cotton fiber.

Cottons having various degrees of acetyl content have been produced for different uses, but those in the range of 17–22% acetyl show good resistance to heat and microbial degradation. Although microscopical studies have been made on many partially acetylated cottons, those in the higher range of acetyl content have been used to demonstrate the effect of excessive heat on the fibers. Congo red stains the fiber a pale pink color but where any damage occurs, the color deepens to a red shade. Fibers in this picture were heated at 160°C for 64 h. Dark areas represent surface damage. Compare with photographs of severely damaged fibers in Plate 5.

Magnification x600

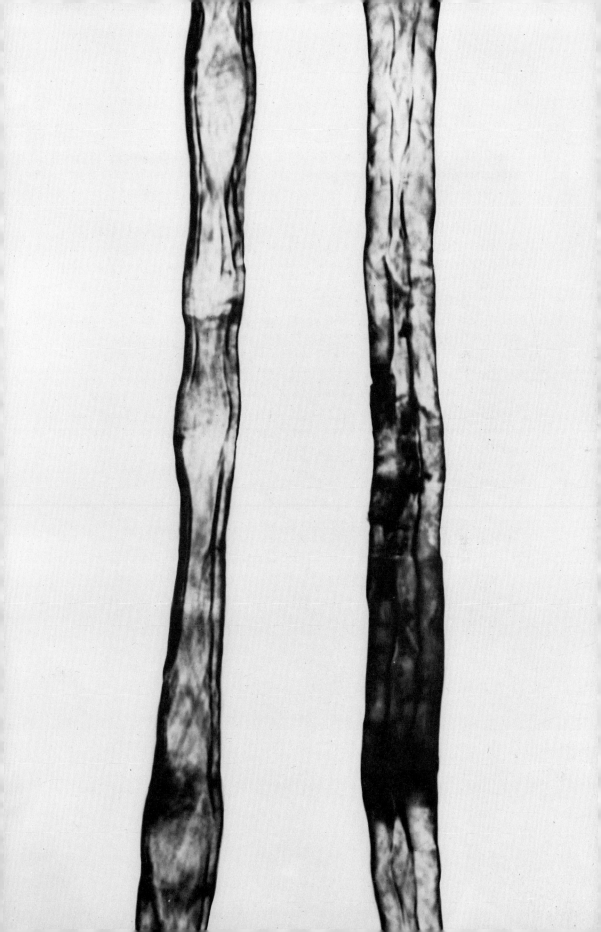

PLATE 7.

Longitudinal view of overbleached fibers showing damaged walls and frayed ends.

Microscopical examination of fibers for effects of damage is of considerable value, not only as a means of detecting the type of damage but as an aid in the improvement of processing to minimize damage.

Chemical tendering from overbleaching results in a gradual breakdown of the cuticle and splitting of the fiber ends. The effects are indistinguishable from those caused by strong acid tendering, except that when Congo red is used the latter shows bright red blotches.

Above all other characteristics of overbleached cotton, reduction in strength is the greatest. This renders the cotton completely useless in a textile fabric, as is demonstrated by the disintegrated fibers in the picture on the opposite page.

Magnification x560

PLATE 8.

Longitudinal view of cut pieces of cotton fiber in 15% sodium hydroxide.

To obtain a picture of this type, short lengths of cotton fibers, about 0.5 mm, are cut from a bundle of fibers. They are then mounted in 15% sodium hydroxide on a slide and examined under the microscope. If the primary wall is free of damage, the pressure of internal swelling forces the secondary wall celluose to exude from either end of the fiber segments while the primary wall remains intact. By counting the number of fibers with this appearance as opposed to those which have been damaged, a quantitative expression of the amount of damage in each sample can be estimated. The test used here is known as the "dumb-bell" test for cotton fiber damage.

Magnification x1000

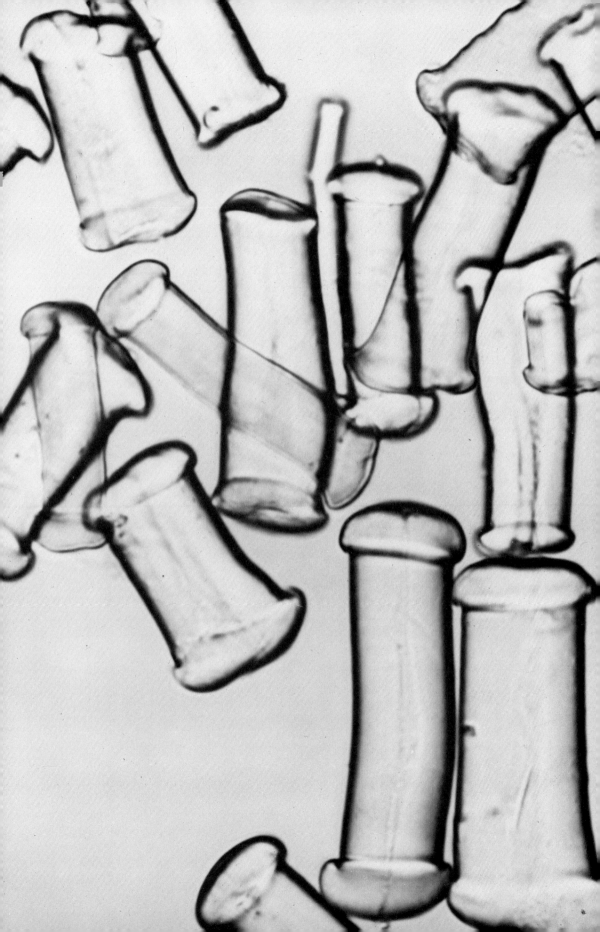

PLATE 9.

Longitudinal view of cut pieces of cotton fibers in 15% sodium hydroxide.

The "dumb-bell" test described on the previous page was used to determine the percentage of damaged fibers in this sample.

When damage to the primary wall has occurred, the wall is unable to restrict the swelling of the secondary wall cellulose caused by exposure to 15% sodium hydroxide. Hence, the entire piece of fiber swells uniformly and the tight collar of primary wall seen in Plate 7 is absent here. By counting the number of fibers of each type on the slide, extent of damage can be estimated on a percentage basis.

Magnification x1125

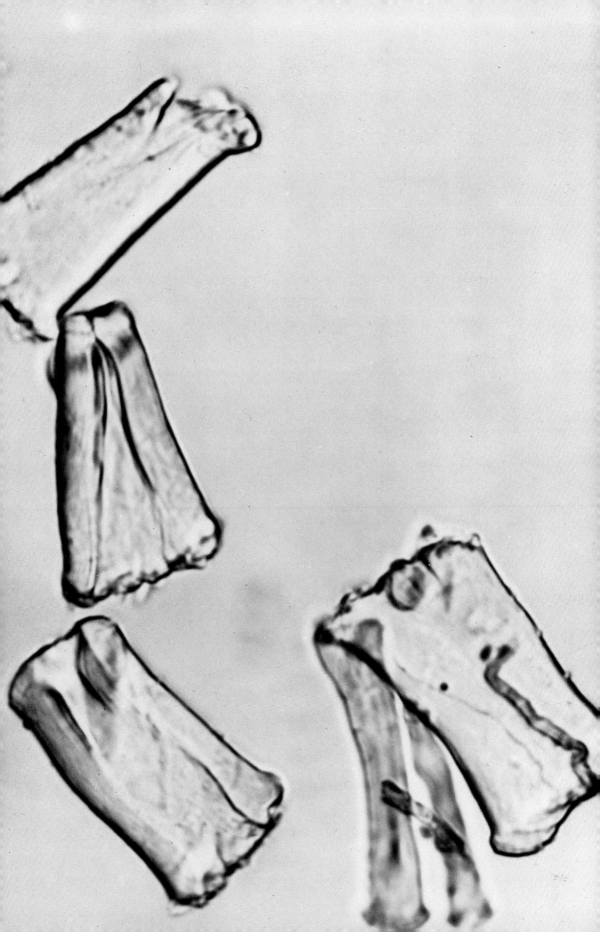

PLATE 10.

Longitudinal view of cotton fibers showing fungal growth.

Under certain conditions of temperature and humidity, cellulolytic organisms grow on cotton. Under the microscope, it is often possible to see the terminal ends of the fungal hyphae and the formation of asci but, to enhance the details of the growth, stains are often used. One of the most popular stains for this purpose is cotton blue in lactophenol which imparts a pale blue color to the cotton and a deep shade of blue or lavender to the fungal hyphae and spores. Besides these formations on the surface of the fiber, it is often possible to see such growth inside the lumen of the fiber. Biological tendering, whether by mildew or bacteria, may resemble chemical tendering, but the disintegrated appearance of the fiber and the presence of hyphae and spores are distinguishing features.

Magnification x560

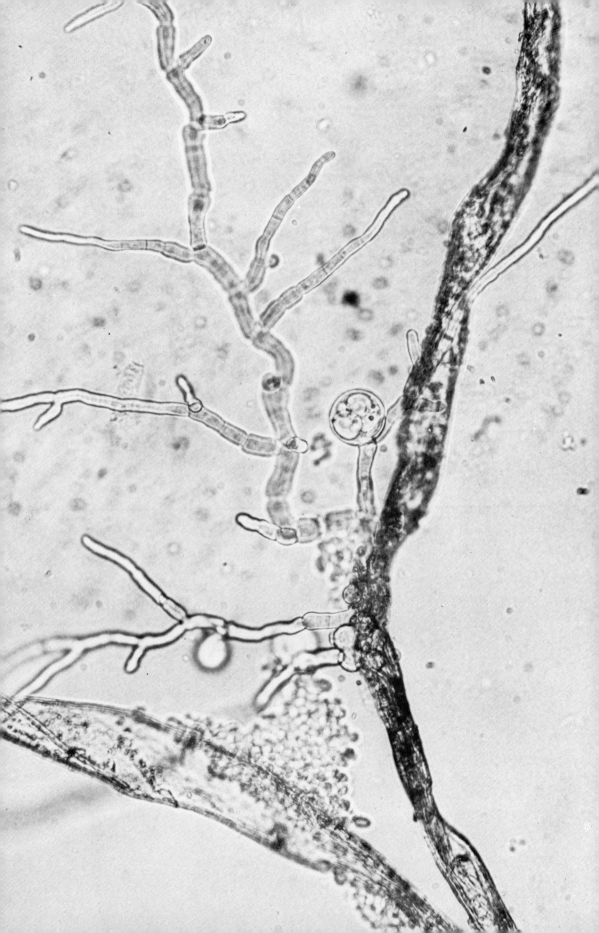

PLATE 11.

Longitudinal view of cotton fibers showing a small sheath of honeydew clinging to primary wall.

A substance deposited on plants as an exudate from aphids or scale insects, honeydew is a type of contamination familiar to most cotton mill operators. Apart from the poor quality yarns it produces, honeydew-infected cotton is even more troublesome than other damaged fibers. Because of its gummy nature, it causes matting, tangled fibers, and stickiness throughout all phases of processing. Simple routine methods are employed to detect the presence of honeydew in individual fibers or in bulk quantities of cotton. Under the light microscope, its natural appearance is that of a transparent, yellow, sheath-like substance. In larger samples, it appears as brick-red clumps of material after exposure to Benedict's copper solution and steam. Unfortunately, however, the mill operator is often not aware of the existence of honeydew in his cotton until machines are affected during processing.

Magnification x685

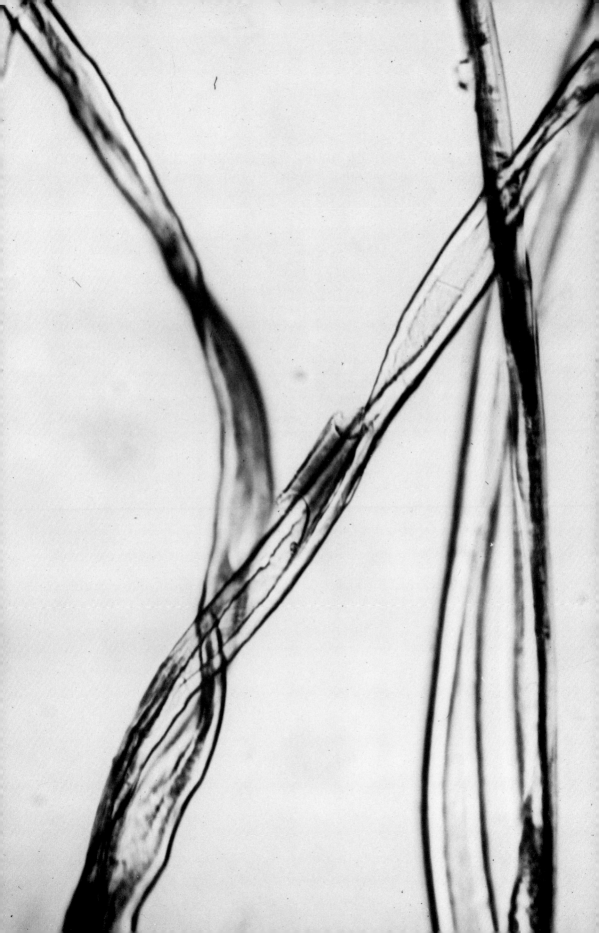

PLATE 12.

Longitudinal view of native cotton fiber showing internal structure after swelling in 0.2 M cupriethylenediamine hydroxide (cuene).

The swelling and dissolution of fibers lend themselves well to microscopical observation. Cotton is dissolved by the copper solvents for cellulose, cadoxen, and solutions of iron tartrate complexes in sodium hydroxide. If swelling is controlled by using a relatively weak concentration of the swelling agent, fiber structure becomes evident. However, in a solution of cupriethylenediamine hydroxide (cuene) of stronger molarity than 0.2 M, osmotic forces set up within the fiber produce writhing, contortion, an increase in diameter, and a decrease in length. These phenomena precede dispersion or solution of the cellulose.

The first thing noticeable in the swelling of cotton fibers in the longitudinal view is the disappearance of all twisting. Following this, the pressure of the swollen secondary wall cellulose causes the primary wall to break and peel back, forming constricting bands. The helicoidal spiralled layer just beneath may then be observed, along with faintly seen secondary wall layers and the twisted contorted lumen.

When a fiber dissolves in a stronger solution, there remain only the residues from the cuticular sheath and the protoplasmic lumen material.

Magnification x500

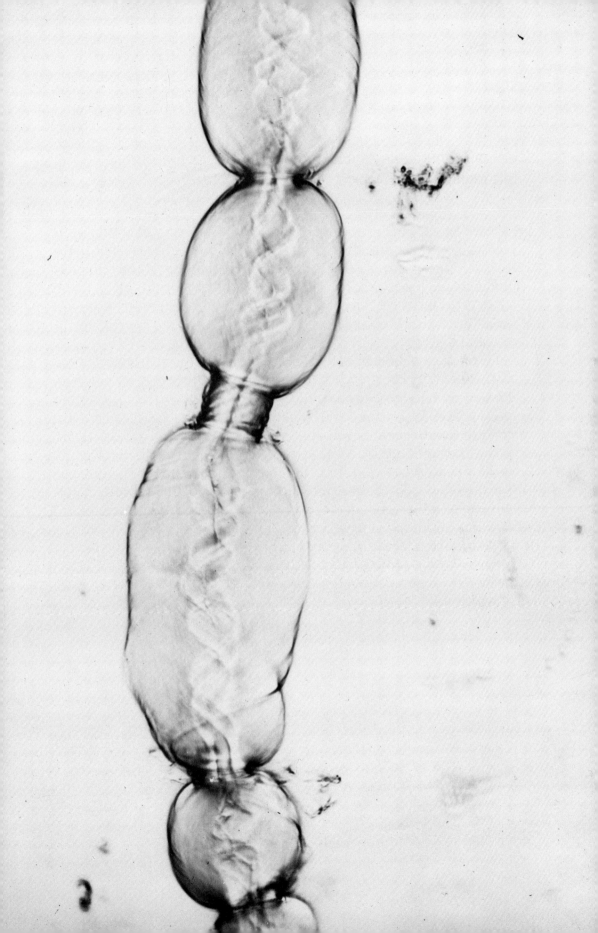

PLATE 13.

Longitudinal view of chemically treated cotton fibers showing limited crosslinking.

Treatments which produce good wrinkle recovery in cotton samples have a direct effect on certain physical properties of the fiber,
such as extensibility, swelling capacity, and solubility. The response
of chemically treated cotton fibers to cellulose solvents such as the
copper-amine bases varies considerably. This phenomenon has
been the basis of an analytical method to evaluate the extent of
crosslinking.

Fibers are soaked in a 0.5 M solution of cupriethylenediamine
hydroxide (cuene) on a glass slide and examined after 30 min.
The extent of lateral swelling of the fibers is indicative of the extent
of crosslinking. In this strength cuene, untreated cotton fibers will
dissolve rapidly; extensively crosslinked fibers will show no response
to this reagent; and fibers of less well crosslinked samples will swell
various amounts depending on the degree of reaction, but will
not dissolve even after extended contact with the solvent. Thus,
comparisons may be made in a series of cottons treated under
varying conditions of time, concentration, catalyst type, and curing
temperature, and samples can be ranked with respect to the degree
of crosslinking on the basis of their swelling response.

Magnification x460

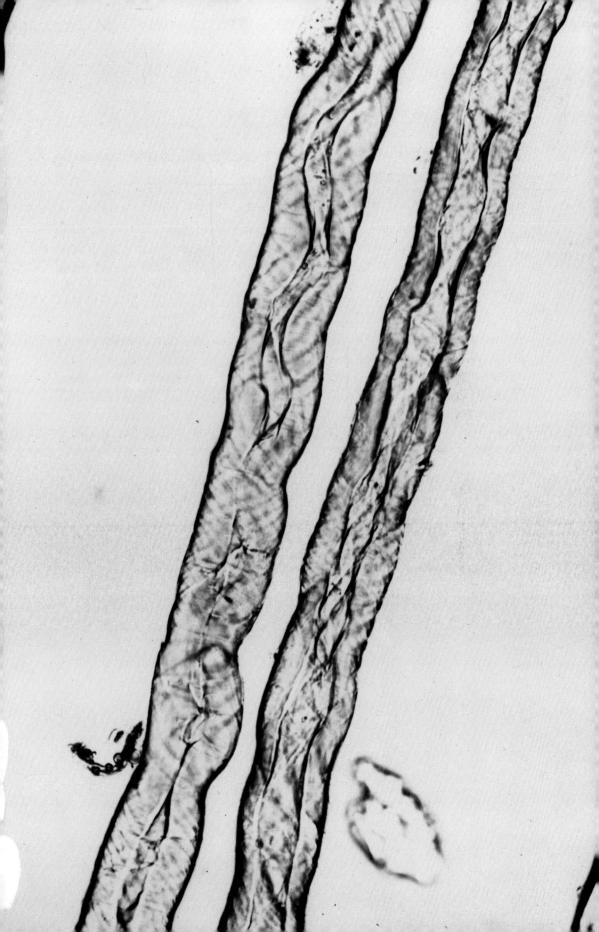

PLATE 14.

Longitudinal view of chemically treated cotton fiber showing extensive crosslinking.

Unreacted cotton dissolves in a 0.5 M solution of cuene, but this chemically treated fiber shows no response, lying immobile in the reagent with no detectable swelling. No features of fiber internal structure are visible and the general appearance of the fiber is similar to that of a fiber of untreated cotton. This lack of solubility in a classical solvent for cellulose indicates that the crosslinking of cellulose chains within the fiber body has been extensive. This extreme degree of crosslinking is rarely observed; the more general case is represented by fibers which undergo some swelling but do not dissolve or break at any point after a 30-min exposure to the cuene.

Magnification x650

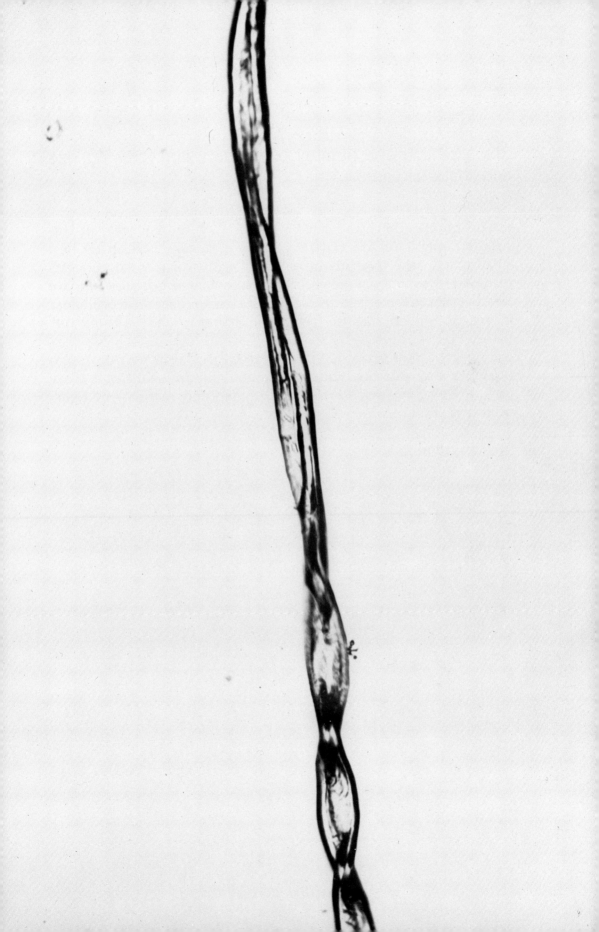

PLATE 15.

Cross section of a number 50s cotton yarn showing size and shape of yarn.

The number here indicates $50 \times 840 = 1$ lb of a yarn of this size. To obtain such a cross section, the sample is inserted in a small hole at the base of the slot in a hand microtome and surrounded by some type of synthetic filaments which hold it in place and prevent mechanical distortion of yarn shape. A nitrocellulose lacquer is applied and the free hand section is made with a razor blade.

Under the microscope it is possible to see that even after thorough blending, drawing, doubling, and combing in the textile mill, there are still extreme variations of size, shape, and maturity of fibers in a yarn cross section.

Magnification x950

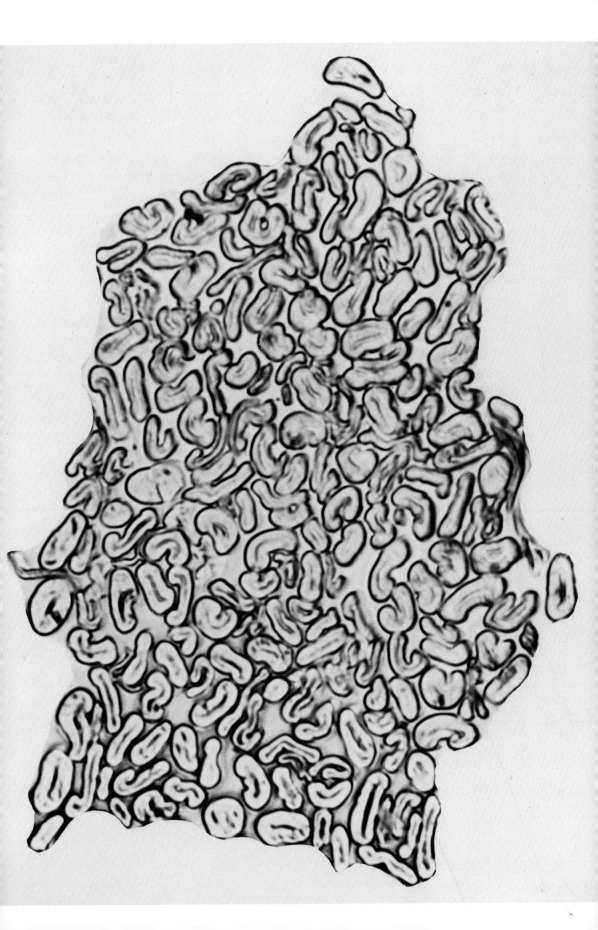

PLATE 16.

Cross section of carboxymethylated cotton yarn dyed with methylene blue to show nonuniformity of treatment.

Partial carboxymethylation produces a quickly swellable cotton fiber which retains its strength and swells when wet. The partial treatment is obtained by the impregnation of cellulose with monochloroacetic acid, followed by a strong sodium hydroxide solution to yield the sodium salt of carboxymethylcellulose. Treated yarns do not vary greatly in breaking strength, feel, appearance, or moisture content from mercerized controls, but swelling capacity and water retention are greatly increased.

Dyeing partially carboxymethylated cotton with aqueous methylene blue results in varying depths of shade, depending on the extent of the treatment. Native cotton dyes pale blue, mercerized a shade deeper, and carboxymethylated a dark blue color. In cross section, these depths of shades indicate the location of the treatment in individual fibers and degree of uniformity of treatment throughout the yarn.

Magnification x450

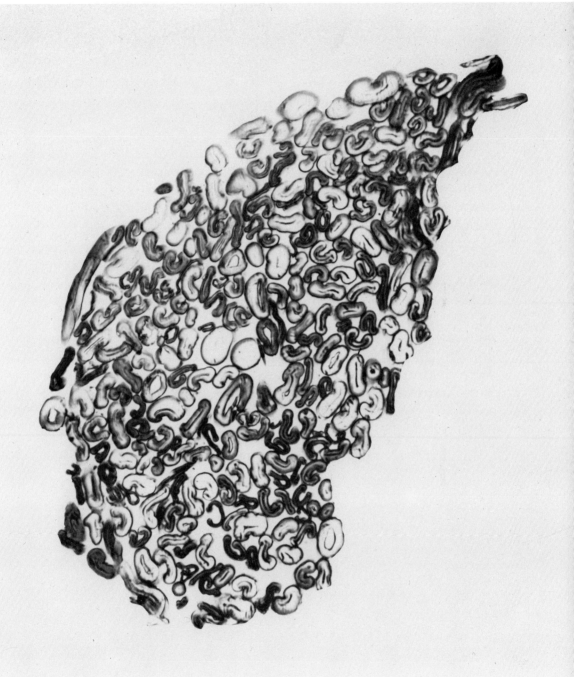

PLATE 17.

Cross section of cyanuric chloride-treated cotton dyed with chlorantine Fast Red 6B11.

The reaction of cyanuric chloride with alkali-treated cellulose gives information on the effect of reaction conditions on degree of substitution and retention of chlorine for further modification. Type of alkali pretreatment and choice of organic solvent appear to influence the reaction.

Changes in dyeing properties were found to be related to degree of substitution in some cases and to type of pretreatment in others.

Microscopical examination of cross sections of cyanuric chloride-treated yarn shows the ring-dyeing of cotton with a dye "built-in" by the chemist. Until the cross section was made, there was no indication as to the location of the dye in the yarn. Throughout this study, the dye appeared as a ring at the outer edge of the fibers, indicating that the dye-forming reaction was only peripheral, and that reagents had not diffused to the fiber interior.

Magnification x450

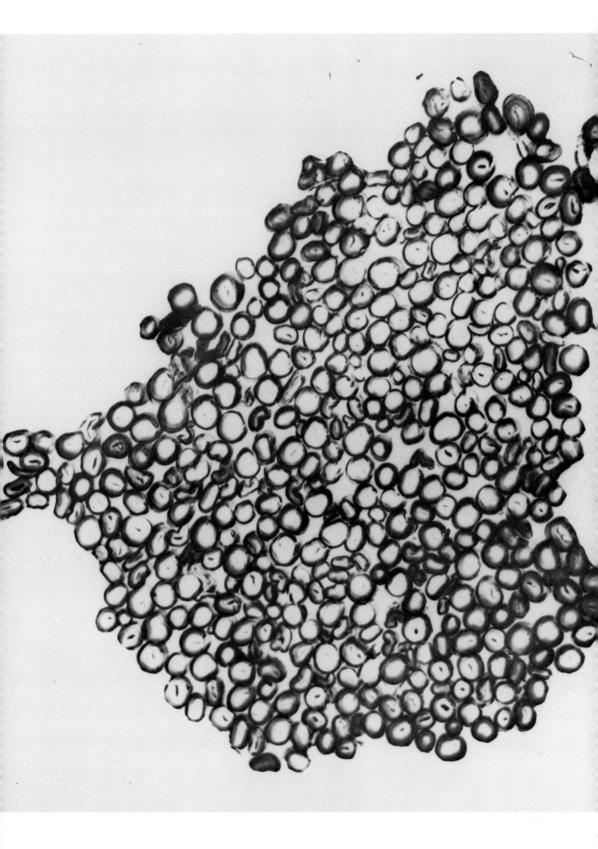

PLATE 18.

Photomacrograph of a cotton fabric showing compactness and degree of uniformity in the weave.

The effects of new developments in textile machinery such as loom attachments can sometimes be demonstrated in pictures of surface views taken through a low-power microscope. With such mechanical attachments, uniformity of weave is improved, the number of reed marks being reduced to a minimum.

Loftiness, a quality which gives added bulk or thickness, can also be observed by this method.

Defects in fabrics are readily demonstrated by examining pictures taken through the microscope. Nep formations, knots composed of masses of textile fibers, are often seen on the surface of fabrics, as are grease stains from mill machines. Other stains caused by bleeding of seed coat fragments or other foreign matter can be determined by examining defective areas.

Precautions can then be taken to improve mill practices, such as proper cleaning of cotton in the early stages and elimination of contaminants from the spinning environment.

Magnification x40

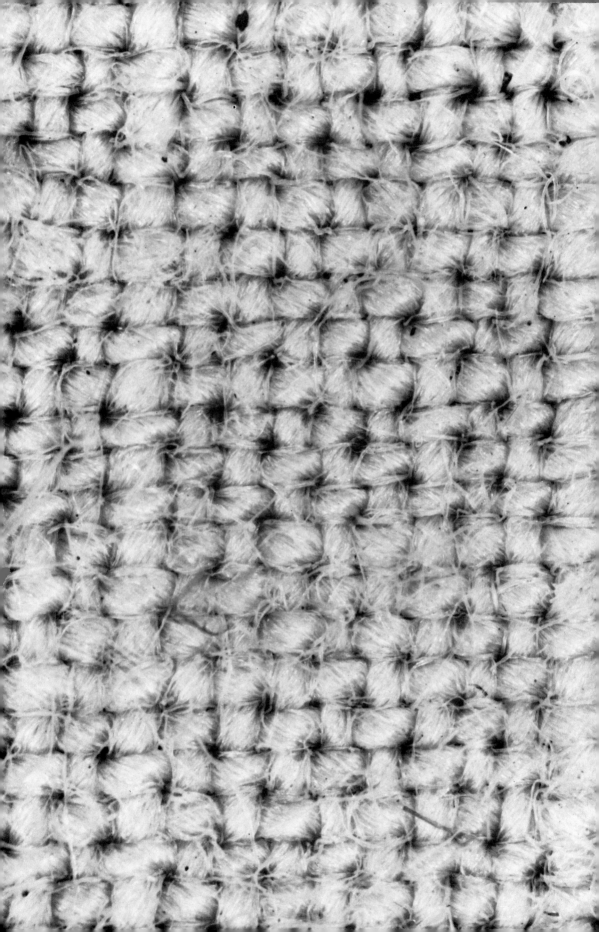

PLATE 19.

Microtome section across the warp and parallel to the filling yarns in a woven fabric.

Information of this type is of particular value to the textile engineer because it is possible to see in detail the denseness of the fabric, the regularity of the yarns, and the angles at which they intersect.

Research on the manufacturer of water-resistant cotton fabrics include development of fabrics of extra high pickage but of lighter weight. These fabrics resist the passage of water largely by virtue of fiber swelling, which results in a closing, or self-sealing effect within the fabric when it is wet. Several factors such as fiber staple length and maturity, and tightness of contact between warp and filling yarn contribute to the effectiveness of this fabric construction.

Magnification x500

PLATE 20.

Microtome section across the filling and parallel to the warp yarns of the fabric seen in Plate 19.

Filling yarns, because of their large diameter, fill the crimp spaces and contribute to the compactness of weave. A construction of this type is valuable in the manufacture of swellable cotton fabrics. Since the average swelling capacity of native cotton is 30%, this, added to the compact weave, would only enhance the closing or self-sealing effect within the fabric when it is needed.

Other fabric properties such as fiber maturity and dye penetration may also be observed in this type of microtome section.

Magnification x500

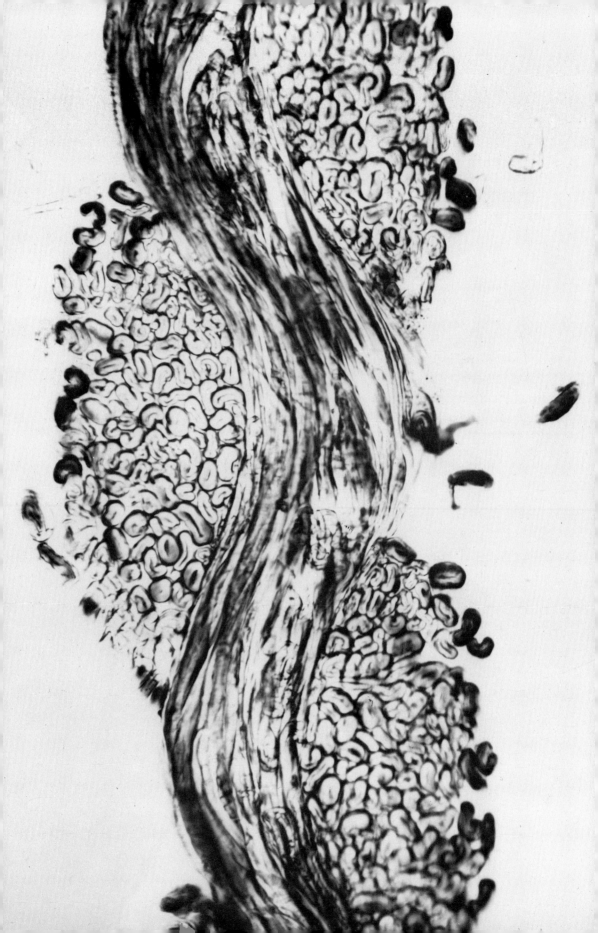

PLATE 21.

Median section through a fabric showing interyarn relationship of warp and filling, and spaces at the interstices of
the yarns.

   This is of significance in the "cover factor" of fabrics designed
to protect against seepage of light or noxious fumes through the
openings in the cloth. Fiber alignment appears to be fairly uniform
in both warp and filling directions.
   To obtain a photomicrograph of this type, the fabric was
embedded flat on a slide in a nitrocellulose lacquer and hand-
sectioned using a razor blade. Such views of yarn contact at midpoint
in a fabric also help to explain differences between fabrics in tests
for air-permeability and water penetration.

Magnification x200

PLATE 22.

Median section through an oxford cloth showing details of fabric construction.

   Minute spaces between fibers in yarns and spaces between yarns in fabrics are regarded as capillary systems. Some of the pores or capillaries spiral along the yarns in the direction of twist. In general, yarns are dense at the center and decrease in density toward the outer edges. As yarns lie in the fabric they tend to press tightly together, but some of the contacts between yarns are through less dense regions, leaving openings at the interstices. Tighter weaving brings about smaller openings, and with water swelling of fibers they become even smaller and result in new capillary systems.

Magnification x225

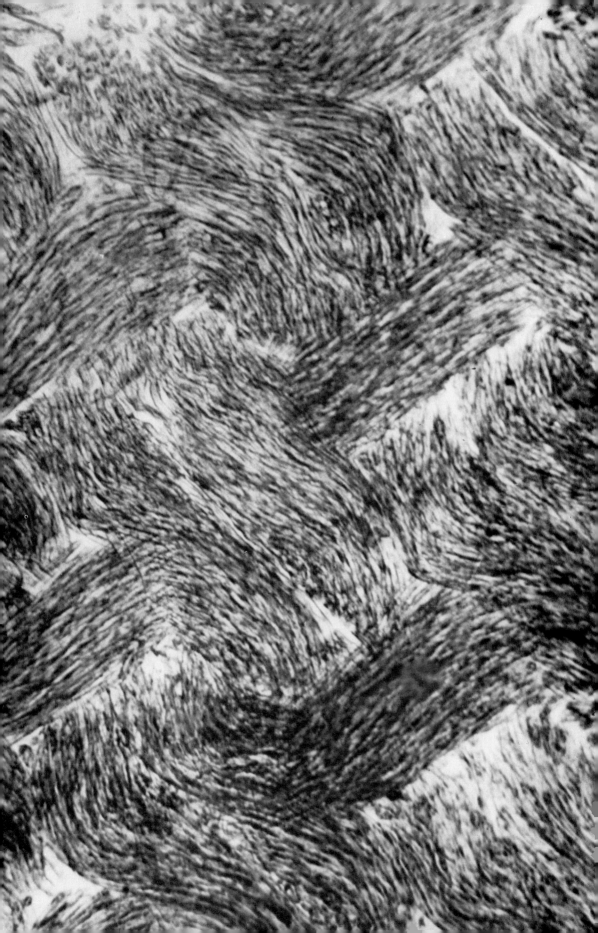

2

TRANSMISSION ELECTRON MICROSCOPY

The transmission electron microscope has been available as a laboratory instrument since 1939, but it was not used in textiles until about five years later. Findings through the years of research with the light microscope have only been enhanced by results obtained with this more complex instrument. Where the light microscope reveals gross structure in fibers, the electron microscope serves to delineate their submicroscopic morphology. Research in this field has resulted in the development of techniques for the investigation of cotton in its native and chemically altered states. Preparation of specimens requires concentrated attention to details to ensure accurate interpretation of electron micrographs.

FRAGMENTATION

When cotton is beaten in water in a laboratory blender the fibers break down into component parts. To obtain pieces of primary wall, the beating process requires only a few minutes; the separation of the winding and secondary layer takes longer.

PRIMARY WALL. Fragments of untreated primary wall generally show little definite structure. Fibers examined prior to the deposition of the secondary wall show some structure, but the mature primary wall shows a fibrillar structure largely masked by the presence of nonfibrillar material. After extraction of wax, pectic materials, and other noncellulosic constituents, the fibrillar network of the wall becomes apparent.

WINDING. In the electron microscope, the winding layer of raw cotton shows a strongly oriented fibrillar texture separated by narrow regions which appear to contain wider fibrils perpendicular to the highly oriented fibrillar groups. Alcohol extraction or scouring more clearly reveals two systems of fibrils: one system of highly parallelized groups, interrupted by a system of coarser fibrils less perfectly arranged.

SECONDARY WALL. Untreated cotton normally yields sheets of cellulose having discrete microfibrils; mercerized fragments show a certain disorientation of microfibrils. Crosslinked cottons usually fracture into relatively short fragments whose microfibrils appear to be fused together in bundles. Such a pattern reflects the brittleness characteristic of the treatment.

SECTIONING

For ultrathin sectioning, cotton fibers are embedded flat on a glass slide in a mixture of poly(methacrylates) methyl-butyl, 3:2, catalyzed by 2% Luperco CDB (50%, 2,4-dichlorobenzoyl peroxide in dibutyl phthalate). Embedding is allowed to harden on a warming table at 65°C for two hours prior to trimming for the microtome. Sections on the order of 900–1500 Å thick are obtained by slicing with a diamond knife beveled to 48°. Subsequently freed of the polymer by vapor extraction with methyl ethyl ketone, they are shadowed with evaporated metal and examined in the microscope.

SWELLING AND DISSOLUTION

Swelling and dissolution tests of sections prepared according to the above method permit assessment of such reactions as crosslinking, substitution, and graft polymerization within the fiber. Cupriethylenediamine hydroxide (cuene) at 0.5 M concentration dissolves unreacted cellulose immediately. Immersion of sections in this solution reveals uniformity of treatment, location of reacted areas, and general fiber behavior.

LAYER EXPANSION

Artificial explosion of biological specimens is the mechanism on which this technique is based. While this ordinarily causes devastating tissue disruption, its controlled use in cotton investigations provides significant information on relative structural changes associated with chemical modification.

When cotton is first boiled in a 50-50 mixture of methanol-water (with 1% wetting agent), and then embedded in butyl and methyl methacrylates, separation of the wet fiber into well-defined layers of relatively unswollen cellulose occurs by rapid linear polymerization of the methacrylates. This is a gross artifact believed to be due to differential rates of polymerization. The wide, open-layered pattern of native cotton expanded in this way is a good basis for interpreting the relative states of bonding between lateral elements within the walls of treated cotton.

REPLICAS

Microtopography of fiber surfaces may be investigated by a two-stage replication process using a viscous mixture of prepolymerized methyl-butyl methacrylates. A negative impression is made in the methacrylate sheet; the final product is a positive carbon impression shadowed with an appropriate metal.

Examination of surface replicas in the electron microscope reveals progressive effects of such processes as scouring, bleaching, impregnation, and coating, as well as physical phenomena associated with damage from abrasion, heat, and microbial degradation.

PLATES

PLATE 23.

Electron micrograph of a sleeve of primary wall removed from cotton fiber by beating in water in a laboratory blender.

The primary wall is the outer skin of the cotton fiber and consists of a network of cellulose microfibrils randomly interlaced and encrusted with noncellulosic materials. By chemical analysis, the primary wall is approximately 50% cellulose, and contains approximately 10% pectic material, 10% fatty material, and 15% proteinaceous material. The cuticle is believed to be a continuous sheet of waxy material covering the whole fiber. The cuticle and the primary wall are usually indistinguishable in microscopical observations.

Often the primary wall is so well covered with noncellulosic material that no fibrillar detail is visible.

In this specimen, many microfibrils can be seen amid the noncellulosic matrix.

The distance between markers represents 1μ.

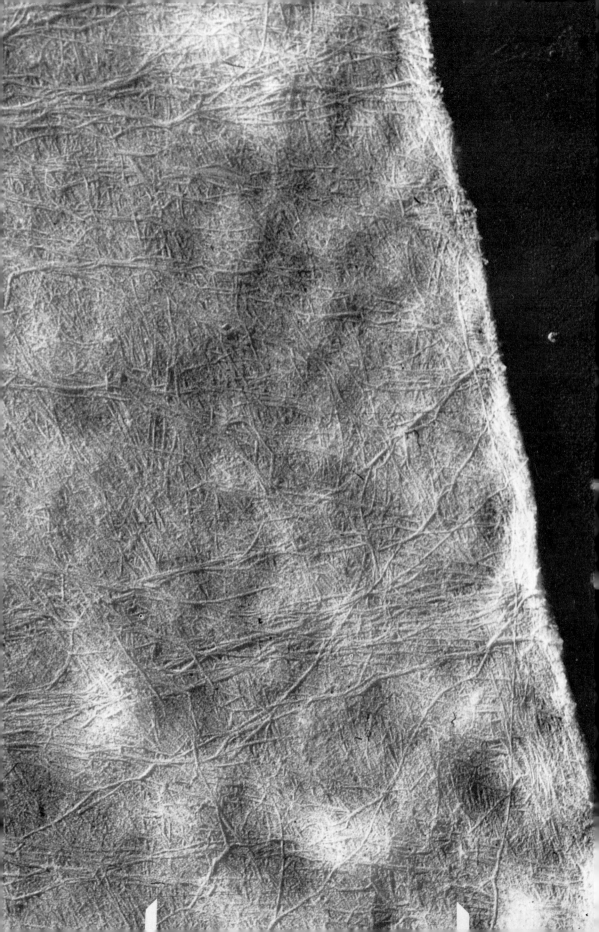

PLATE 24.

Electron micrograph of a primary wall fragment after purification treatments.

The removal of noncellulosic materials by appropriate extraction unmasks the fibrillar structure of the primary wall. Fibrils on the outer face of the specimen appear to lie more or less parallel to the fiber axis; those at the lower level lie at various angles to the fiber axis.

It has been shown that both number and diameter of microfibrils in the primary wall increase during the growth period. Electron micrographs of a purified primary wall from a two-day old fiber showed very fine microfibrils, much less numerous and less linearly oriented on the surface. It has been suggested that the arrangement parallel to the axis is the result of tensions developed during the longitudinal growth of the fiber.

Upon swelling in chemical reagents, individual microfibrils increase in diameter and shorten. It is obvious from this photograph that the net effect of this phenomenon is shrinkage of the primary wall sleeve into a constricting casing around the fiber.

The cellulose primary wall network shown here is not removed from the fiber in most treatment processes. In fact, after commercial-type kiering and bleaching, its fibrillar structure is not visible, being still obscured by the noncellulosic materials which are only partially removed by normal textile finishing purification.

PLATE 25.

Winding layer of the cotton fiber.

The first layer of secondary wall deposited immediately inside the netlike primary wall of the cotton fiber is referred to as the "S-1 layer" or "winding layer" of the fiber wall. It serves as a transition layer between the network of the primary wall and the oriented fibrils which make up the subsequent layers of the secondary wall in the main body of the fiber. Its structure is that of a wide system of bands or tapes lying at a wide angle to the axis of the fiber. Two systems of microfibrils are seen here—the more solid on the left and the open "lacy" network on the right. The alternation of these two systems of microfibrils is irregular; the wider bands have a closely packed parallel arrangement of fibrils while the narrower bands (which connect them) consist of two systems of coarser fibrils at right angles to one another in an open "leno-weave" pattern. The alignment of this fibrillar system is 45–70° to the fiber axis.

The winding is usually seen only in fibers which have suffered chemical or microbial damage, but is often observed as a residual "skin" after dissolution of the fiber in cellulose solvents.

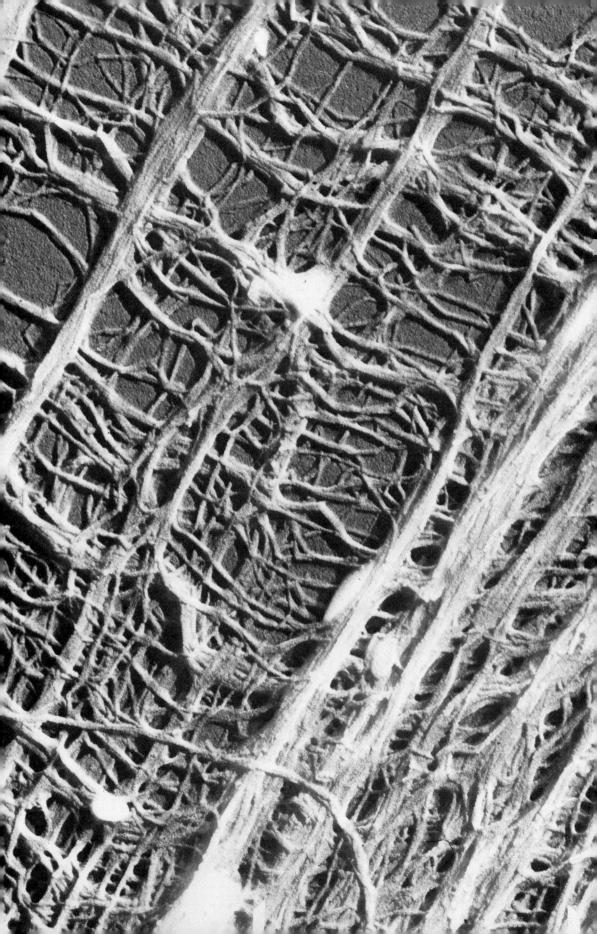

PLATE 26.

Fragment of secondary wall.

The main body of the cotton fiber consists of cellulose microfibrils arranged in a more or less parallel array, as pictured here. Fragments such as these are isolated by beating fibers in water in a laboratory blender. Microfibrils are closely bound in a parallel system which appears to be continuous; there is no interruption of the pattern, as in the bonded structure of the winding layer.

Current concepts of cellulose structure indicate that microfibrils such as those pictured here are aggregates of finer cellulose threads, "elementary fibrils," whose cross-sectional measurements are approximately 35Å. The elementary fibril is considered to be a brittle, needlelike crystal on the faces of which regions of slight disorder have cellulose chains available for reaction. Along the length of the crystal, occasional regions of disorder exist as a result of slip planes, faults, and other dislocations occurring during disposition and growth. It is at the elementary fibril level of fiber structure that the chemical modifications of textile finishing must take place.

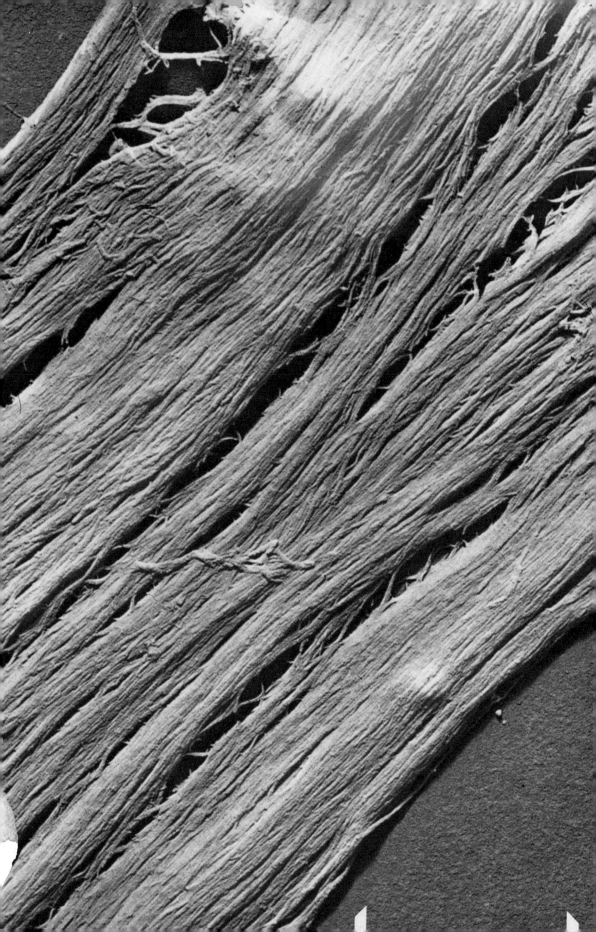

PLATE 27.

Fragment of chemically substituted cotton.

Chemical substitution of the OH groups in the cellulose chain permits conversion of the cellulose. Thus, by reaction of cotton with acid chlorides of monobasic acids, partial esterification of the cellulose can be achieved, retaining the natural fibrous properties of the cotton and imparting additional properties of the derivative. By partial acetylation, for instance, improvements can be made in stability to heat, and in resistance to moisture and microbial deterioration.

The transmission electron microscopy of fragments produced in the disintegration of fibers by beating in water in a laboratory blender has shown that untreated cotton breaks apart within 5 min into long sheets or strings of sharply defined fibrils. Modified cottons require longer periods of beating for breakdown into fibrillar components; the decreased water absorption of the esterified cellulose is probably responsible for this increased difficulty of fibrillation. The resulting fiber fragments are characteristically different: at low degrees of substitution, the fibrillar appearance of fragments differs little from that of fragments from untreated cotton, but as degree of substitution (D.S.) increases, the fragments appear as a mixture of fibrillar material and amorphous, gummy lumps.

This fragment of cotton fiber secondary wall from a sample with 8% acetyl content indicates that a D.S. of 0.32 acetyl groups per anhydroglucose unit is too low to cause much change in the physical appearance from that of the untreated secondary thickening.

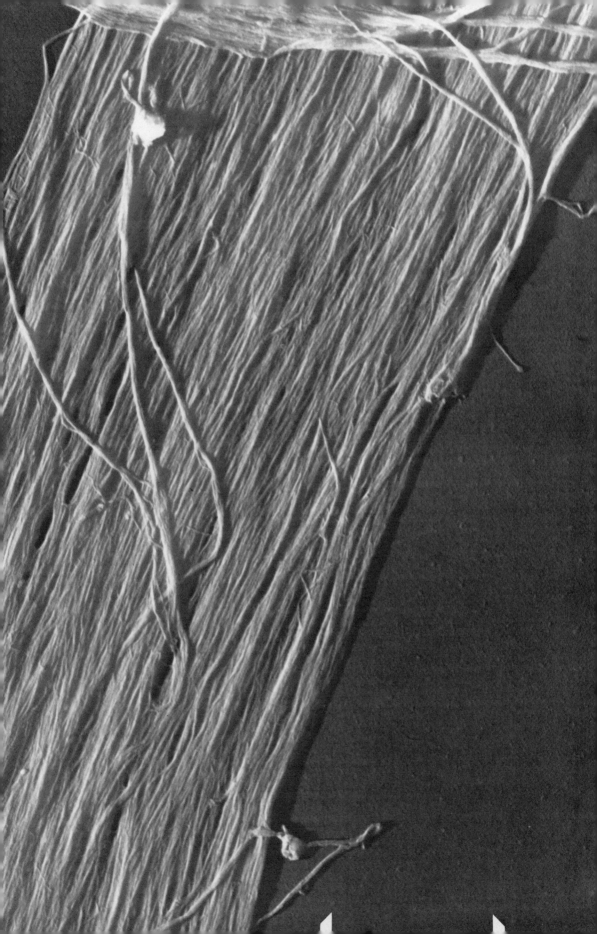

PLATE 28.

Transmission electron micrograph of a fragment of acetylated cotton (D.S. = 0.94).

Fragmentation in water in a laboratory blender of partial cellulose esters produced from cotton yielded long strands of fibrillate material intermingled with clumps of spongy or amorphous material. The higher the degree of substitution (D.S.), the more gummy aggregates are formed on beating and the fewer discrete fibrils are observed. This specimen, obtained by beating cotton with a 20% acetyl content, shows an apparent fusion of fibrillar components within the sheet, and shows both brittle fracture and plastic agglutination of portions of the fiber fragment. It is obvious that a D.S. of 0.94 is high enough to cause a change in the fibrillation pattern with considerable fiber fusion.

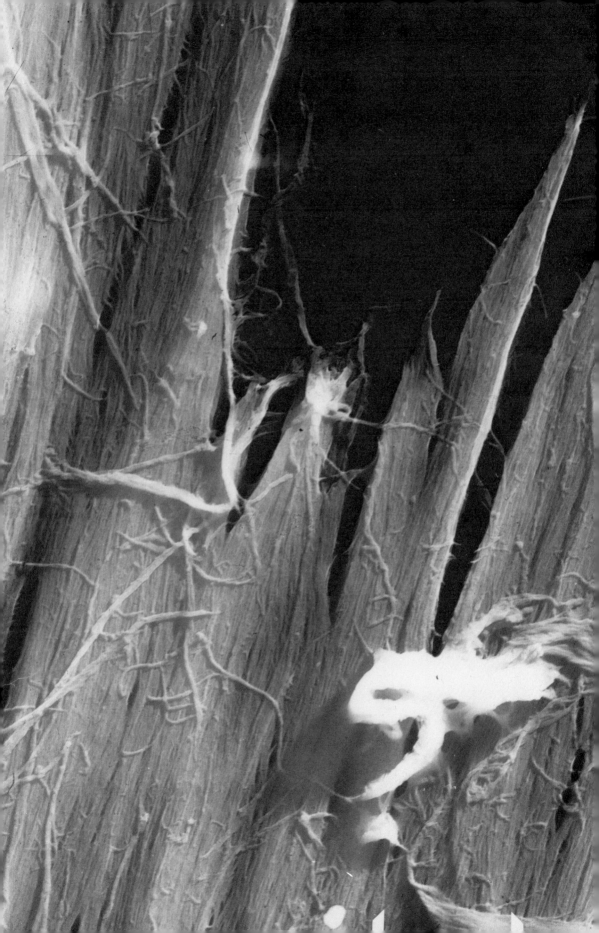

PLATE 29.

Effects of partial esterification on the structure of the cotton fiber.

In the comminution by wet-beating of cellulose esters of cotton of higher D.S. (degree of substitution), the amorphous component predominated in all fields examined. This specimen of the secondary wall of cotton containing 38.5% acetyl is composed almost entirely of clumps of spongy material. Thus, at a D.S. of 2.32 acetyl groups per anhydroglucose unit of cellulose, the fibrillate character of the water-beaten slurry is virtually eliminated. Loss of fibrillate character can conceivably be brought about by plastic flow during beating. The overall temperature during the procedure used in these studies rose to approximately 50°C. Transient temperatures in localized regions of the specimen would have been much higher due to impact, and thermoplasticity would have permitted flow in those regions of the specimen susceptible to high temperature. Fragments of acetate rayon (secondary acetate) and of Arnel (cellulose triacetate) produced by wet-beating possess little fibrillate character and resemble the photograph here presented. On the other hand, fragments of regenerated cellulose saponified acetate fibers were more fibrillate than amorphous.

The effects of partial esterification on the morphology and submicroscopic structure of cotton appear to be readily detectable by microscopical methods.

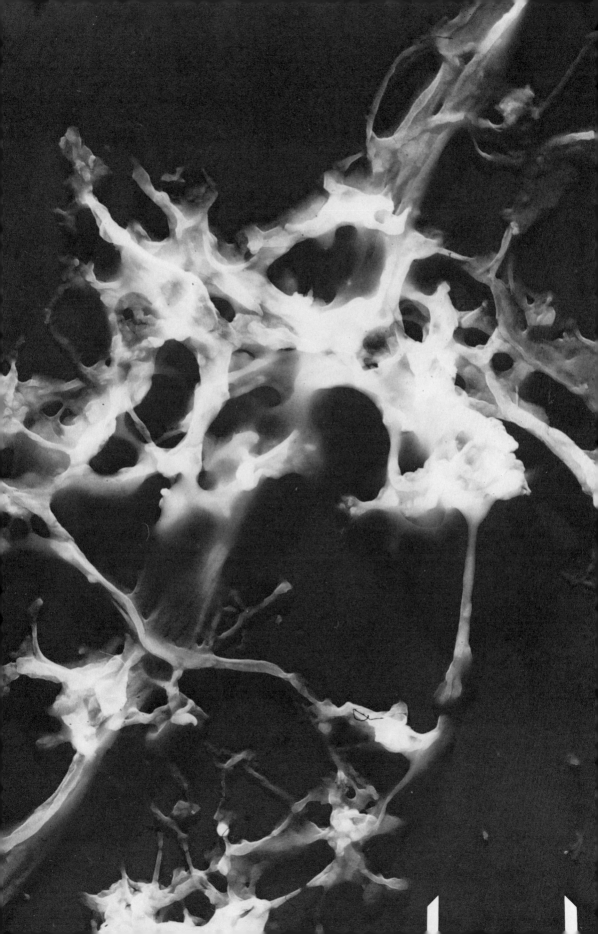

PLATE 30.

Fragment of crosslinked cotton showing a typical sharp break in fiber.

Chemical modifications to achieve wrinkle-resistance or to improve dimensional stability of fabrics involve the possibility of chemical linking of the cellulose to the impregnating reagent, or the formation of crosslinking bonds between the cellulose members themselves within the fine structure of the fiber. The effect of these changes can be observed microscopically. Upon beating in water, cotton samples treated with certain crosslinking agents yield fragments consisting of short fiber segments with little fibrillation; few of the long, stringy sheets characteristic of untreated cotton are present. The fragments themselves are compact, truncated, and jagged, and most of them consist of two or more layers of cell wall which adhere together tightly. The appearance of fibrillar shearing and fracture is dependent upon the amount and type of crosslinking which may have occurred within the fiber structure.

This fragment from a fabric crosslinked by treatment with dimethylol ethyleneurea (DMEU) shows both primary and second-ary wall layers welded together, and indicates the brittle nature of the fiber after crosslinking. The fine trash which accompanies the larger fragment is also typical of fragmentation patterns of highly crosslinked specimens.

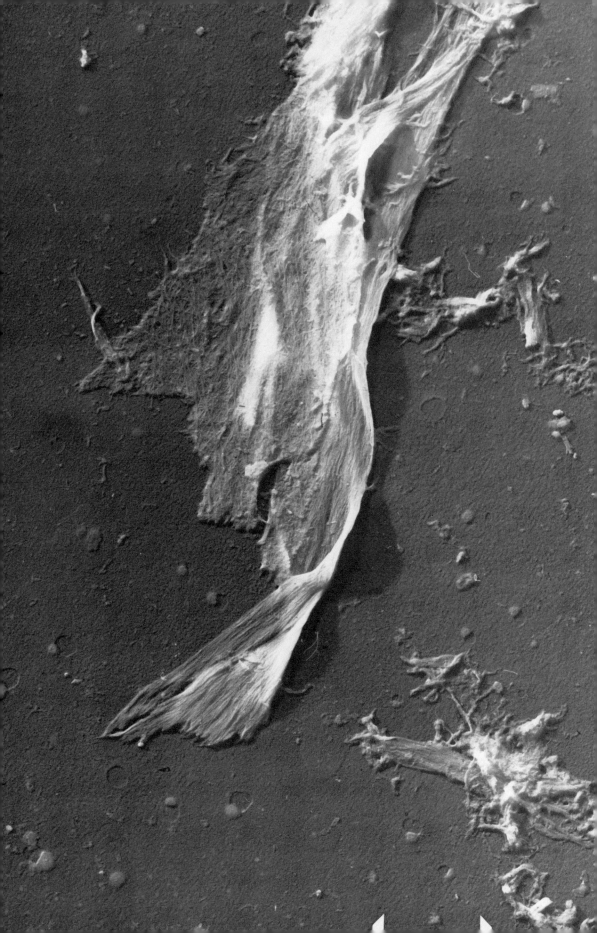

PLATE 31.

Fragment of cotton showing the effect of extensive cross-linking.

Durable-press features of cotton apparel fabrics are largely dependent on covalent crosslinking of the cellulose in cotton, or hydrogen bonds between cellulose chains. Details of the microfibrillate arrangement of the cellulose of the fiber interior are revealed by electron microscopical observation of fragments obtained by disintegration of the fiber in water. Crosslinked fibers yield few long sheets of discrete microfibrils, as observed in untreated cotton, but yield many shorter fragments with compact structure and sheared ends, indicating brittle fracture. The individual microfibrils are very difficult to discern, appearing to have been fused together.

The fragmentation patterns of rafts of microfibrils, indicating abrupt cracking off from the main fiber body, reflect the brittleness which accompanies extensive crosslinking, the restriction of freedom of movement of fibrillar elements, the reduced modulus of rupture, the loss in breaking strength, the decrease in extensibility, and the low abrasion resistance so often observed in durable-press cotton fabrics.

Intensive research to diminish these effects while retaining the wrinkle-resistance features have led to very satisfactory compromises which have made such treatments commercially successful.

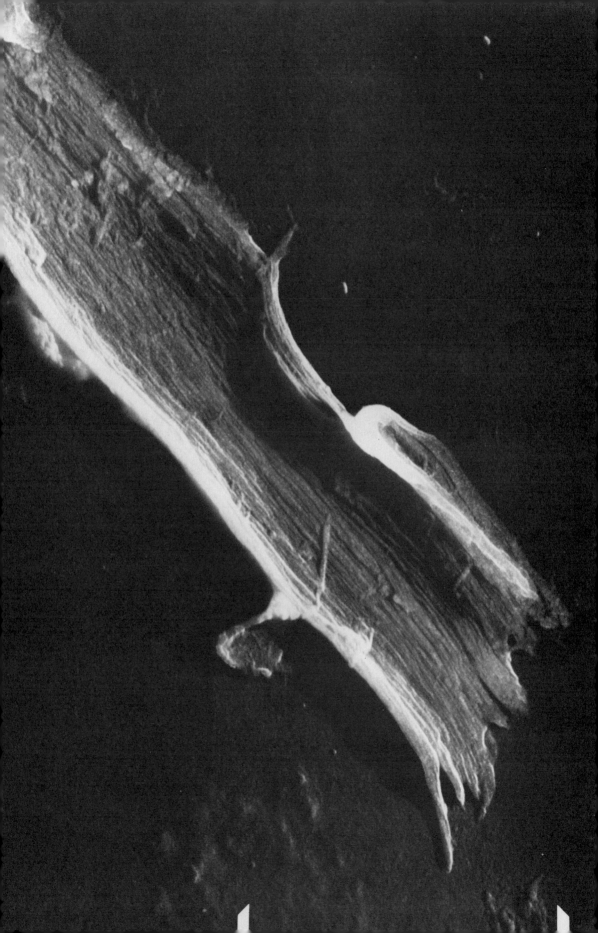

PLATE 32.

Fragment of amine-decrystallized cotton fiber.

The structure of this fragment of cotton derived by wet-beating of a sample which had been partially decrystallized by amine treatment appears to be very little different from that of conventionally scoured cotton. However, X-ray measurements indicate that the degree of order has been reduced, the percentage of highly ordered crystalline cellullose is decreased, and, thus, the accessibility to liquids has been increased. This circumstance enhances dyeability and improves chemical reactivity.

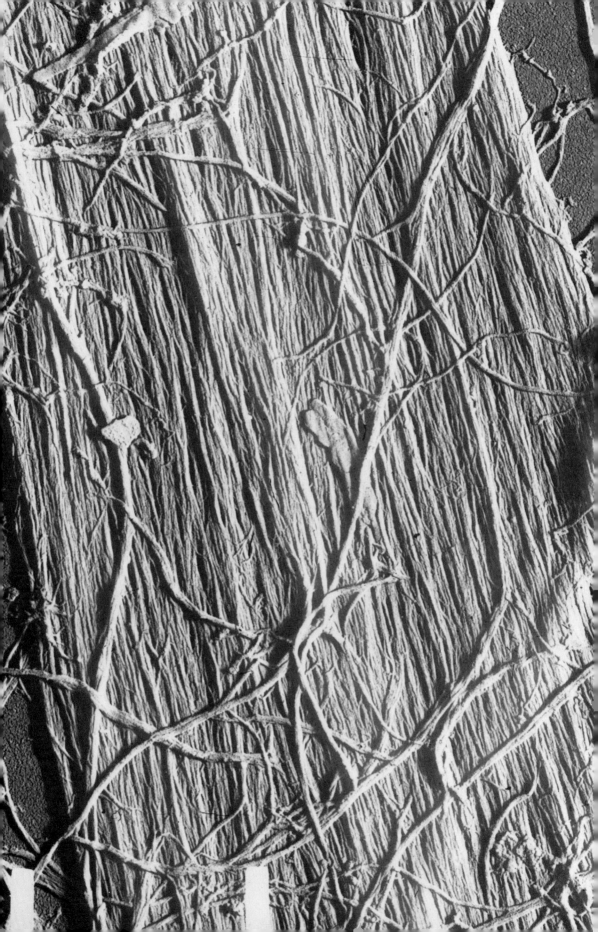

PLATE 33.

Fragment of slack-mercerized cotton.

This specimen, obtained by wet-beating of cotton which had been mercerized in the slack state (i.e., without restraint to shrinkage, and without added tension) shows microfibrils with noticeable crimp. Mercerization involves immersion of the yarn or fabric in a sodium hydroxide solution of approximately 20% concentration and results in swelling of the width of the microfibrils and shortening in their length. It is accompanied by disruption of the crystalline nature of the cellulose to bring out a permanent change in fiber properties. Mercerized yarns have improved luster, enhanced sorption and dyeability, and increased strength.

If tension is applied while the fiber is in the swollen, plasticized state, the crimping so obvious in this specimen is pulled out and microfibrils become somewhat fused in a position more nearly approaching straight alignment with the fiber axis.

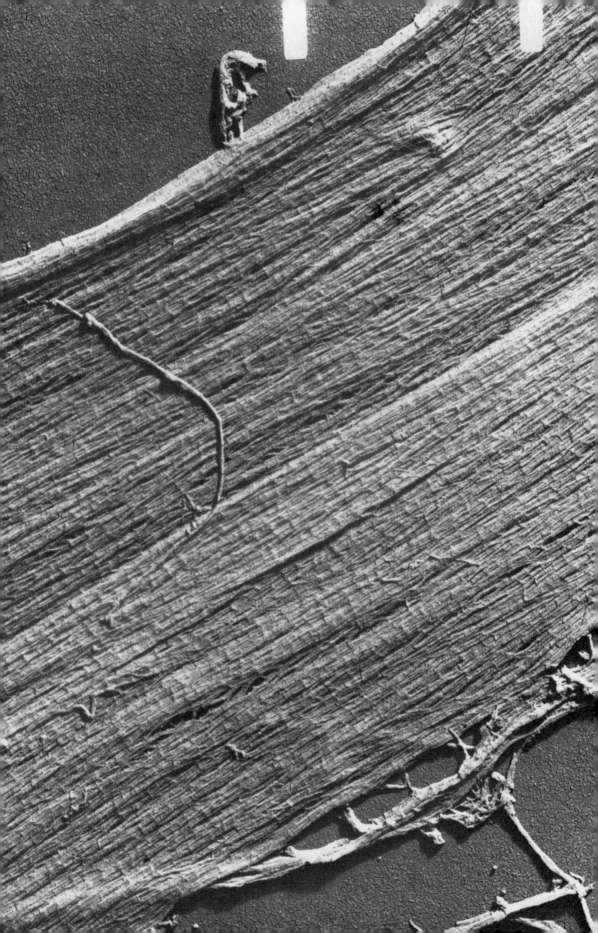

PLATE 34.

Ultrathin cross section of cotton fiber after treatment with dimethylol ethyleneurea (DMEU) for permanent press properties.

At low magnifications many fibers, either treated or untreated, show little or no structure. There is, however, a barely discernible difference between the main body of the fiber and its thin outer sheath, or primary wall.

    This fiber was embedded dry in methyl-butyl methacrylate by polymerization. After sectioning, the slice was freed of the embedding material by gentle extraction with methyl ethyl ketone. The extracted section was shadowed by vacuum evaporation of platinum.

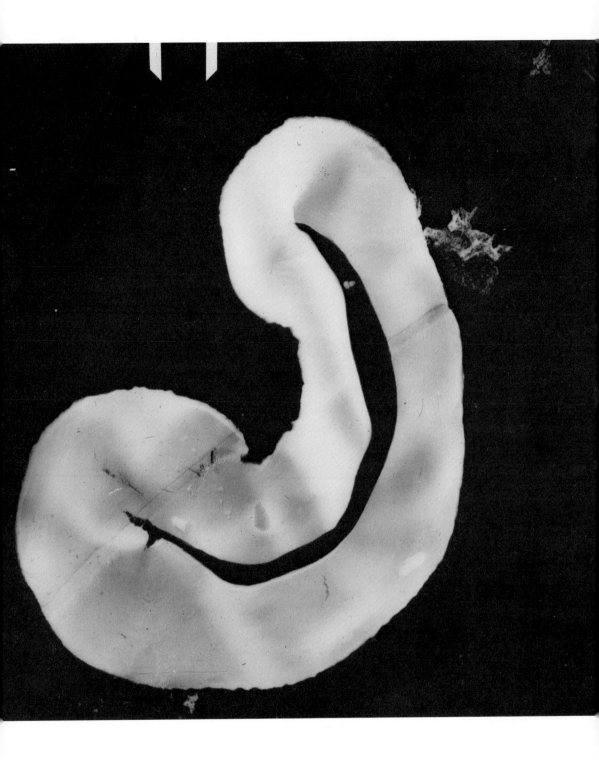

PLATE 35.

Ultrathin cross section of fiber in Plate 34 photographed at a higher magnification.

Ultrathin sections viewed at elevated magnifications can often reveal details of structure useful in comparison of chemical treatments. This section appears to have a somewhat granular structure in the main body of the fiber, which may indicate clusters of cellulose microfibrils held together by induced crosslinking.

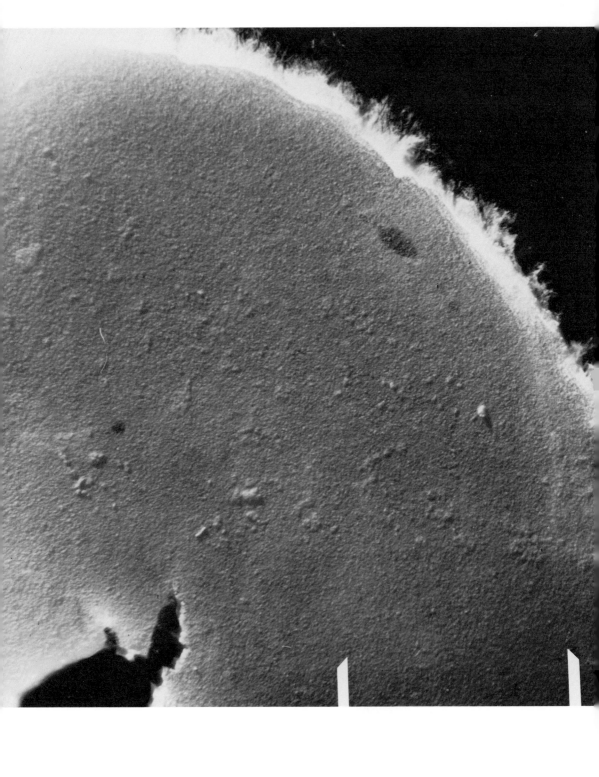

PLATE 36.

Effect of cellulose solvent on ultrathin cross section of cotton fiber treated with dimethylol ethyleneurea (DMEU).

After sectioning and removal of embedding methacrylate, the section was exposed for 30 min to a 0.5 M solution of cupriethylenediamine hydroxide (cuene). This particular reagent dissolves unreacted cellulose; the "lacy" network seen in this electron micrograph is presumed to indicate a limited degree of crosslinking between cellulose units in the submicroscopic structure of the fiber. Wrinkle-recovery features of the fabric would be better than those of untreated cotton, but not as satisfactory as those of a fabric in which the cellulose was extensively crosslinked.

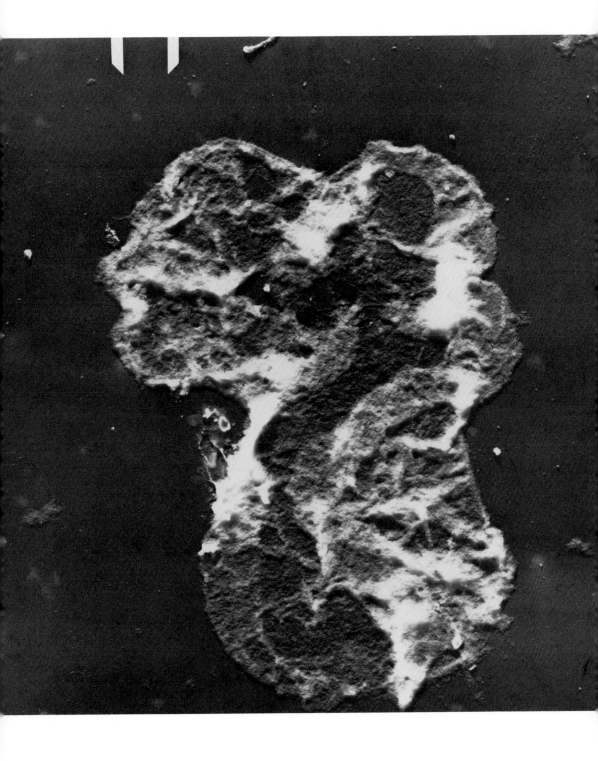

PLATE 37.

Enlarged view of the section shown in Plate 36.

Ultrathin cross section of a cotton fiber with limited degree of crosslinking as indicated by microsolubility in cuene. It is obvious that much of the cellulose has dissolved away, and that in the remaining areas of the section the cellulose is highly swollen.

PLATE 38.

Effect of cuene on microsolubility of an ultrathin cross section of cotton fiber treated with dimethylol dihydroxy-ethyleneurea (DMDHEU), using $Zn(NO_3)_2$ as catalyst.

This section exhibits extensive crosslinking as shown by the amount of reacted cellulose remaining after exposure of the fiber to cuene. Although the section appears somewhat thinner, there are no breaks in continuity and reaction is uniform throughout the section. It is to be expected that wrinkle-recovery characteristics of the fabric from this treatment would be excellent.

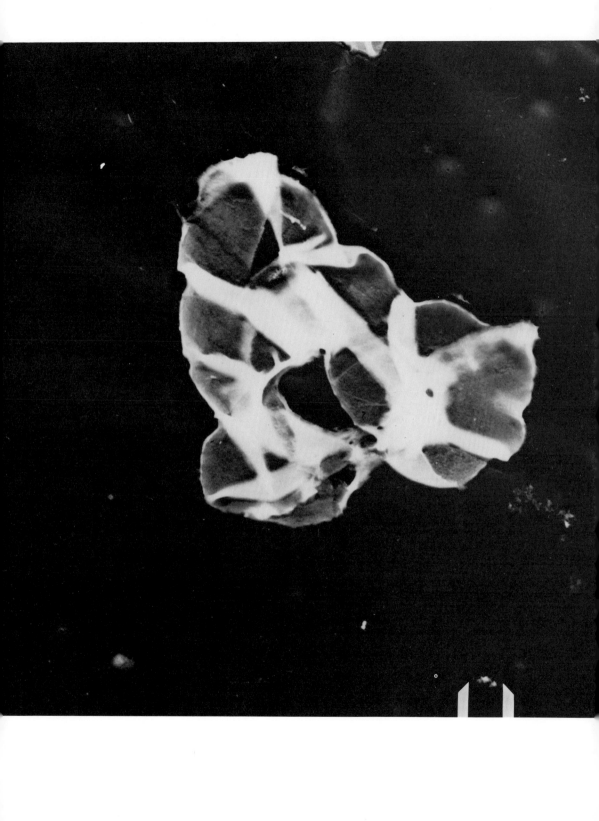

PLATE 39.

Enlarged view of cross section shown in Plate 38.

Ultrathin cross section of a cotton fiber crosslinked by dimethylol dihydroxyethyleneurea (DMDHEU) ($Zn(NO_3)_2$ catalyst) after 30-min immersion in cuene.

The texture of this cross section, which is revealed at this higher magnification, appears to be solid and considerably less granular than that of the fiber seen in Plate 37.

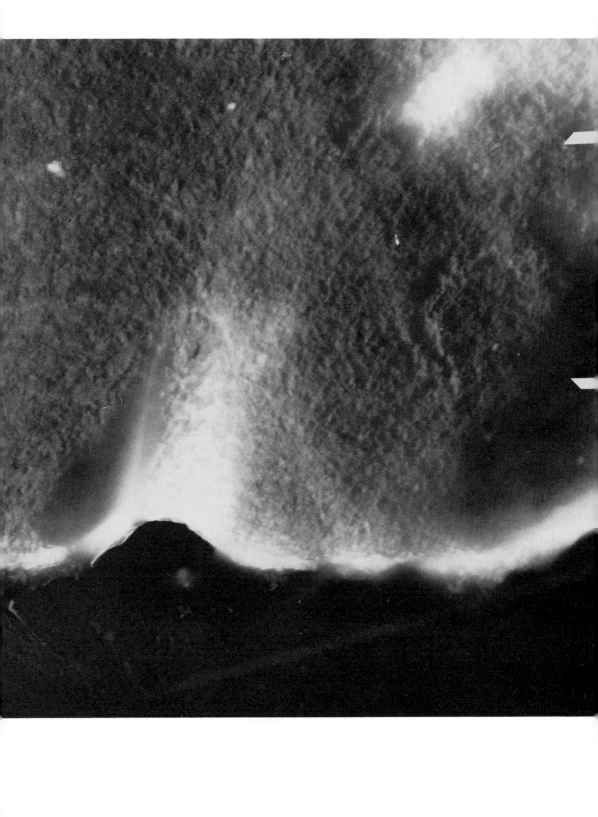

PLATE 40.

Ultrathin cross section of a cotton fiber showing the effects of graft polymerization with methyl methacrylate.

The core of the fiber appears to be unreacted. The reacted outer area has been swollen and partially dissolved by the methacrylate employed in embedding for sectioning, and by the solvent used for removal of the embedding polymer.

Grafting was initiated by radiation of the cellulose with gamma rays from a cobalt source prior to contact with the polymer solution. Adjustment of concentrations of the polymer in its solvent would be expected to improve uniformity of this graft.

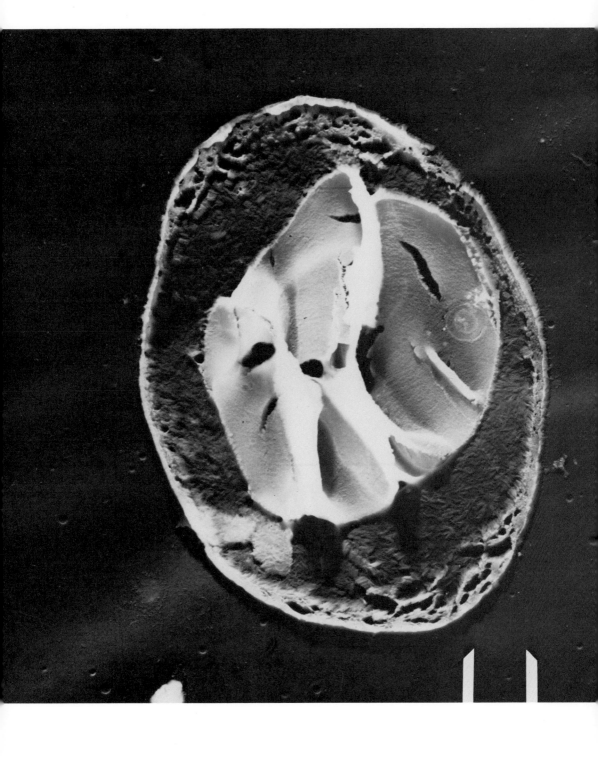

PLATE 41.

Enlarged view of ultrathin section shown in Plate 40.

Two entirely different textures are visible here: the fused texture of the swollen copolymer of cellulose and methacrylate, and the rigid texture of the unreacted cellulose of the core of the fiber.

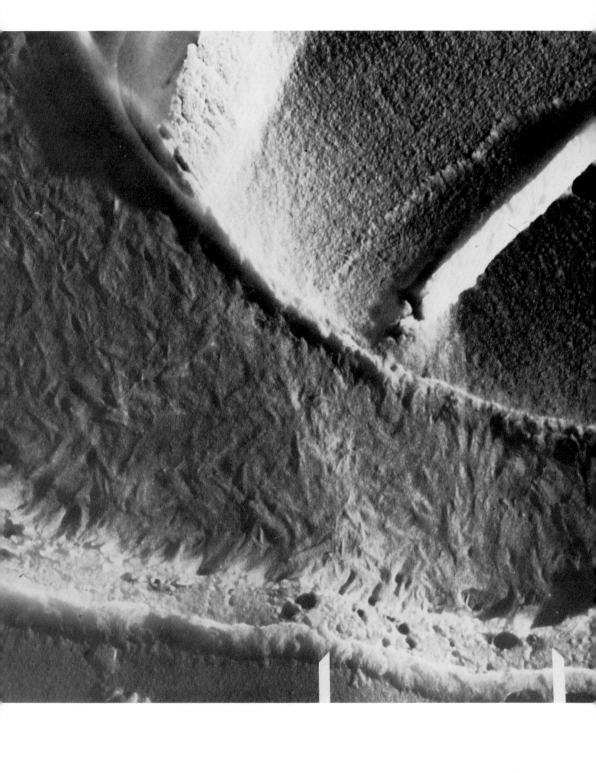

PLATE 42.

Ultrathin cross section of graft-polymer of cotton and methyl methacrylate after immersion of the section for 30 min in cuene.

The unreacted cellulose has dissolved out of the center, leaving the copolymer as a residue at the outer edge. This dissolution of the main body of the fiber by the cellulose solvent confirms the supposition that it was unaffected by the grafting reaction.

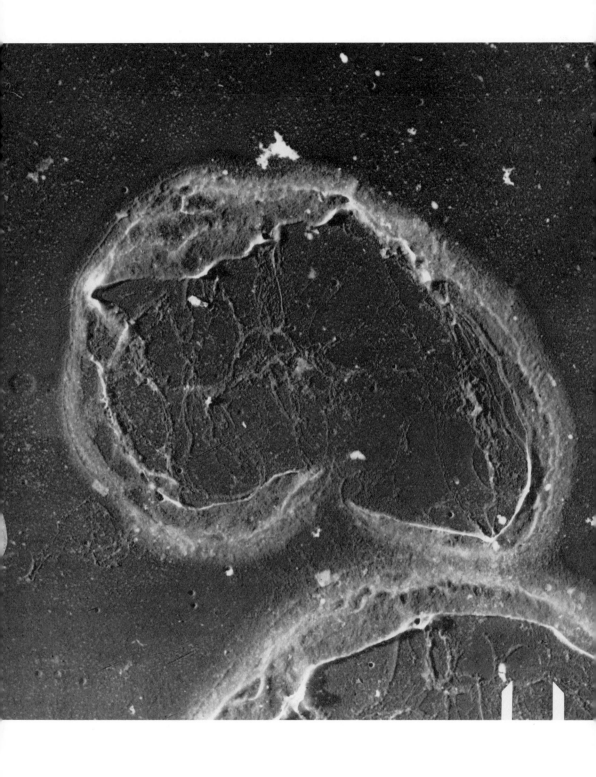

PLATE 43.

Enlarged view of the ultrathin section of Plate 42.

Effect of cuene microsolubility is shown in this section of cotton graft-polymerized with methyl methacrylate by high energy irradiation.

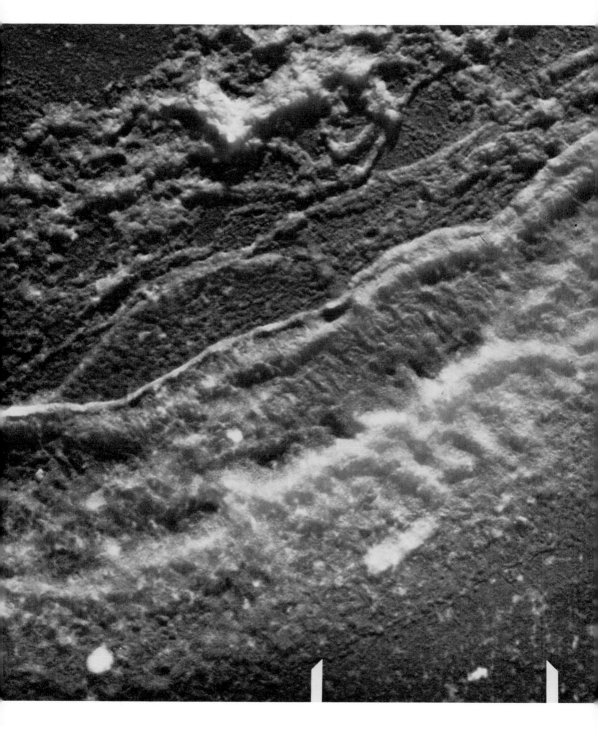

PLATE 44.

Peripheral crosslinking in a fiber from a cotton fabric treated with bis(methoxymethyl)ethyleneurea (BMMEU).

This ultrathin cross section has been immersed in cuene for 30 min. Unreacted cellulose has thus been removed by dissolution. The solid material remaining is presumed to be sufficiently cross-linked so that it is insoluble in the reagent which readily dissolves unreacted cellulose. Since this material is only in the peripheral regions of the fiber, one concludes that further penetration of the reagents into the fiber was in some way impeded. It is possible that crosslinking occurred so rapidly as to set up a barrier of cross-linked cellulose in the peripheral region which effectively prevented further diffusion of reagents.

Such information on the locus of reaction is obtainable only through microscopical techniques, but is often essential in development of new formulations for the chemical modification of cotton to meet end-use requirements.

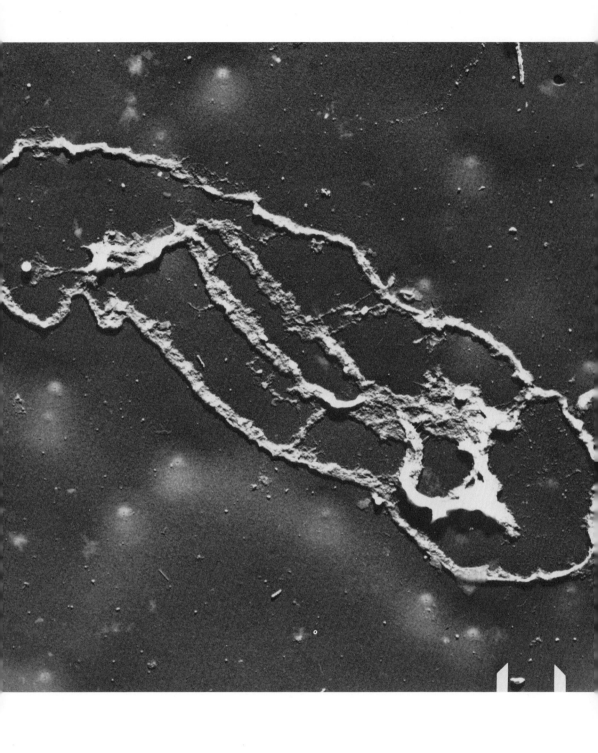

PLATE 45.

Enlarged view of ultrathin section of Figure 44: peripheral crosslinking with BMMEU.

That the peripheral area of the fiber did not dissolve in the cellulose solvent implies that it was extensively crosslinked, or that it was a copolymer of cellulose and the ethyleneurea derivative.

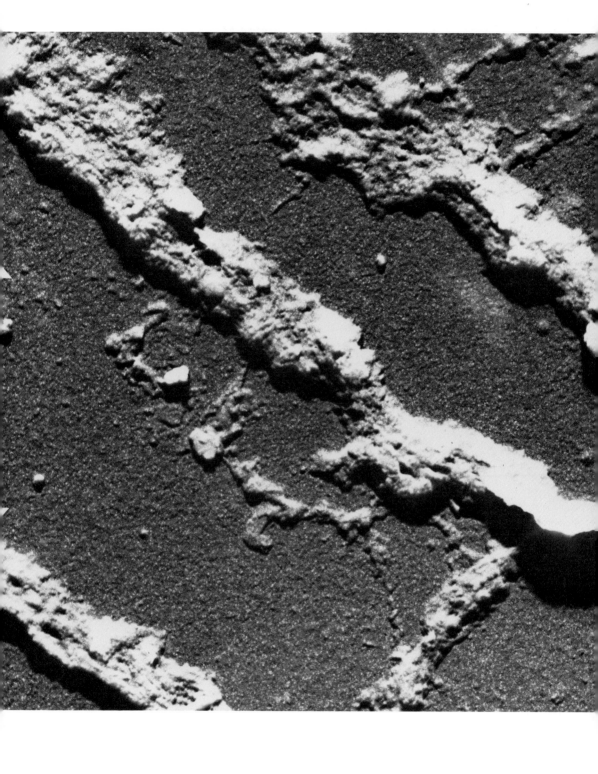

PLATE 46.

Cross section of untreated cotton fiber prepared by the layer-expansion technique.

As a result of the rapid polymerization of butyl methacrylate which has penetrated into the natural cleavage areas of the fiber enlarged by swelling in 50% aqueous methanol, the layers of cellulose within the fiber wall are forced apart. In untreated cotton, the layering is typically extensive and uniform, and, when the embedding polymer is dissolved out, the unsupported layers often collapse sufficiently so that they present a lateral surface view.

In fibers from fabrics given crosslinking treatments to enhance high wrinkle recovery characteristics, no layering occurs.

In untreated fibers embedded dry in methacrylate, no layering occurs.

Variations in the layering pattern on embedment in the wet state reflect any variations in swelling in the prehistory of the sample.

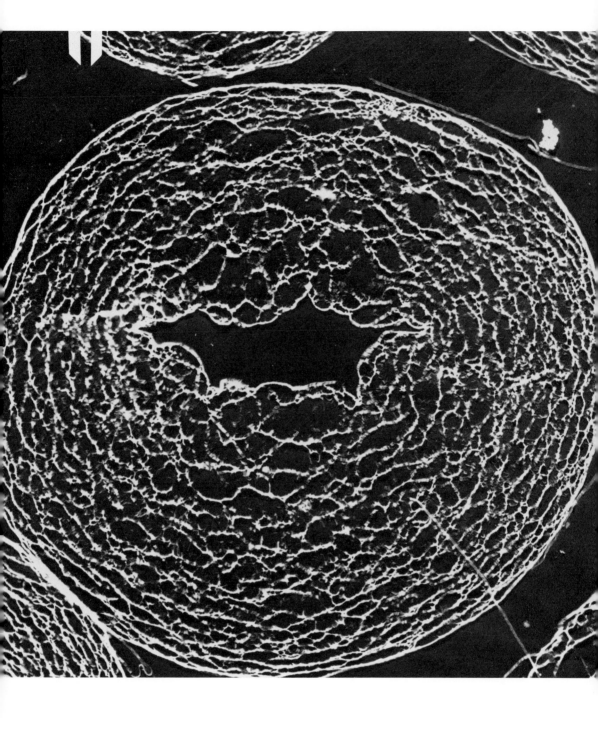

PLATE 47.

Enlarged view of the ultrathin section of Plate 46: layer-expansion of untreated cotton fiber.

In the "tipped-over" layers of cellulose within the fiber wall one may observe the microfibrillar pattern of the cellulose of which each layer is composed. It is presumed that in the treatments used to produce wrinkle-resistant, durable-press fabrics, the microfibrils are covalently bonded one to the other so that moisture cannot be absorbed in interstices between microfibrils, or onto the faces of elementary fibrils which compose them.

Thus, response of the fiber to the layer expansion technique reflects changes in water-accessibility and can be used effectively in comparisons of samples in a series of treatments to screen out significant samples for further evaluation.

Under specific conditions of swelling, untreated cotton fibers and those which have been crosslinked to only a limited extent can be induced to separate into layers of fibrillar material. Under identical swelling conditions, cotton fibers which have been cross-linked to a greater degree do not form layers and, in fact, show no change in cross-sectional appearance. These phenomena, though not in themselves proof of crosslinking, provide useful means of comparison of experimental treatments.

PLATE 48.

Layer expansion pattern of ultrathin cross section of cotton fiber treated with dihydroxyethyleneurea (DHEU).

Ineffectiveness of the crosslinking treatment is indicated by the layer separation; this fiber, not unlike the native control seen in Plate 3, shows poor crosslinking as determined by the expansion method.

Although concentricity of layers is less pronounced in this specimen than is usually observed in untreated cotton, it is still obvious that bonds between layers of microfibrils were slight.

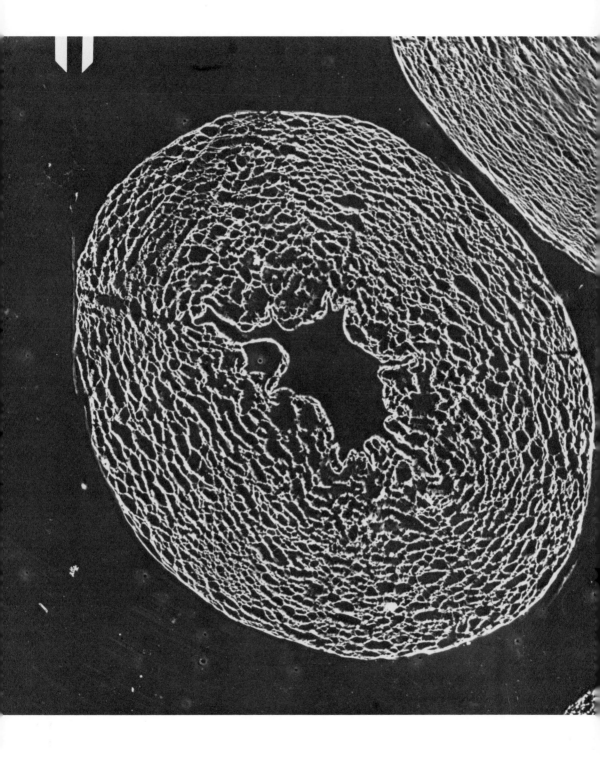

PLATE 49.

Enlarged view of portions of cross section of Plate 48.

The layers appear to anastomose fairly regularly, but fibrillar detail in individual layers is lacking. In fact, it would appear that considerable fusion and some dissolution may have occurred during the reaction.

PLATE 50.

Ultrathin section of cotton fiber treated with dimethylform-amide (DMF) and dimethylol ethyleneurea (DMEU), showing partial layering by the methacrylate expansion method.

Compared with similar expansion pictures of untreated cotton, the layering in this section is less than discrete. On the other hand, the fact that layering occurs at all implies a low level of crosslinking between lamellae.

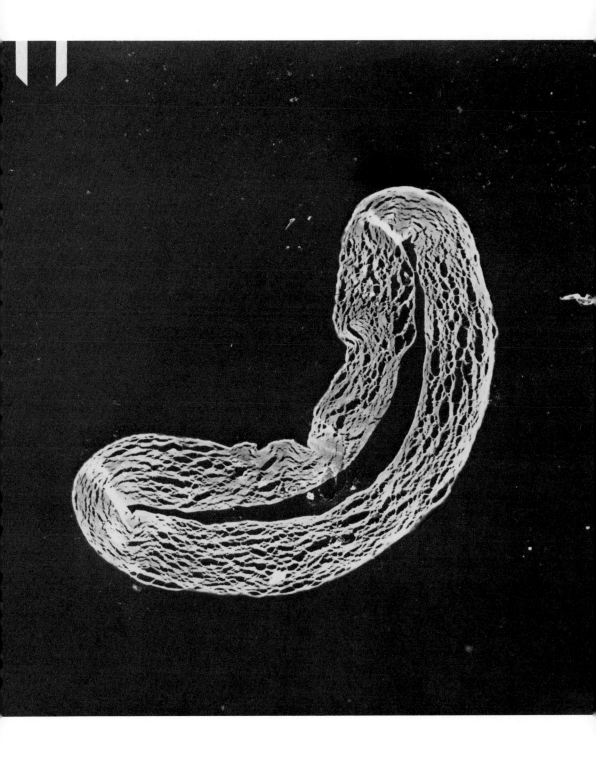

PLATE 51.

High-magnification view of the fiber section in Plate 50, showing a fused appearance of layers.

Although some separation of layers occurred, their component parts are not discrete. It would appear that lamellae are bonded together by crosslinks into fairly thick layers, but that the reaction was not uniform throughout all layers in the fiber wall.

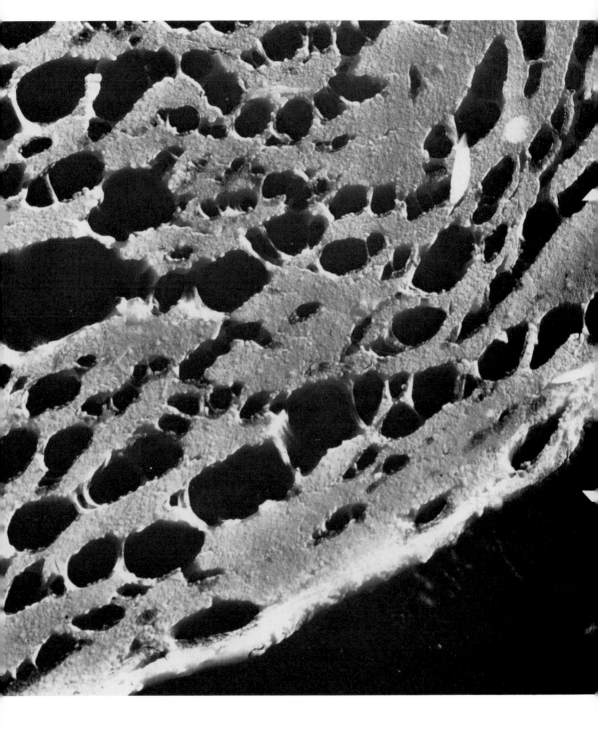

PLATE 52.

Ultrathin section showing layer-expansion pattern of a fiber of diethylaminoethyl (DEAE) cotton treated with epithiirane.

The outside edge of the fiber shows a solid band of what appears to be fused cellulose; the core of the fiber has layered in a manner resembling that of native cellulose. It is assumed that early reaction in the fiber periphery set up a barrier to diffusion, preventing reactants from reaching the center regions of the fiber.

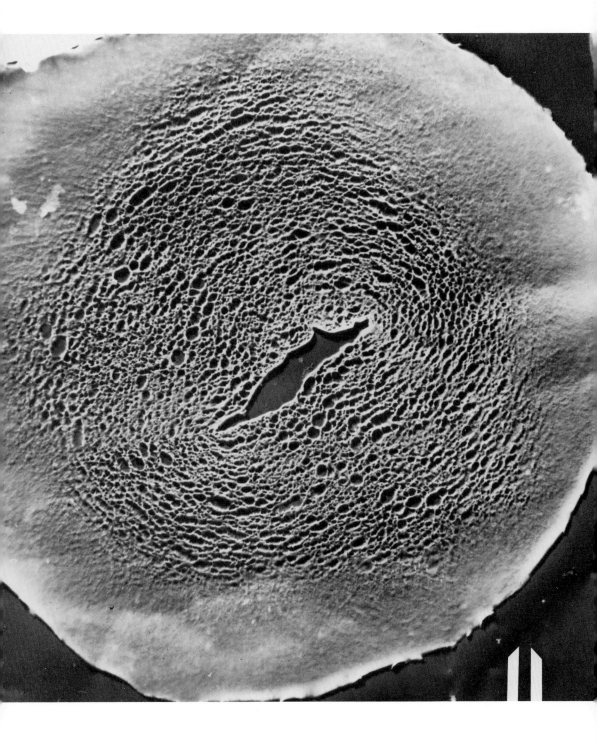

PLATE 53.

High-magnification view of the outer portion of the fiber section of Plate 52.

The outermost edge shows solid polymer of DEAE-epithiirane-cotton cellulose. Further in, the texture is granular, becoming less and less compact as the unreacted center of the fiber is approached.

PLATE 54.

Ultrathin section of a cotton fiber practically intact after swelling according to the water-methacrylate expansion technique.

This fiber exhibits unusually good crosslinking imparted by treatment with dimethylol dihydroxyethyleneurea (DMDHEU) applied in a moist-cure process. The fabric was dried to 10–12% water on a tenter frame and batched for 21 h at room temperature.

PLATE 55.

Enlarged view of outer area of section from Plate 54.

The solid appearance of the cell wall is contrasted with the granular texture often observed in sections of fibers from other types of treatments. It is of interest that the primary wall region of the fiber is more electron dense than the main body of the fiber.

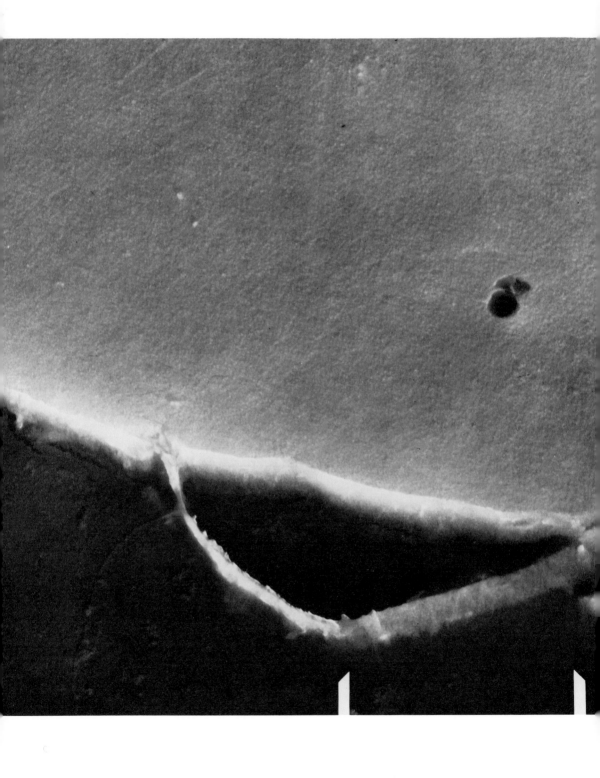

PLATE 56.

Ultrathin section showing a layer expansion pattern of a heat-damaged cotton fiber.

The sample was exposed for 1 h, 15 min to 200°C; broken edges and fissures extending deeper into the fiber show deterioration resulting from excessive heat over a long period of time.

Before cotton reaches the consumer as a textile product, the fibers are subjected to many natural, mechanical, and chemical situations that damage fiber surfaces and change fiber properties. Rarely are temperatures used in drying cotton at the gin high enough to cause fiber damage, but the drying of chemically treated fabrics may occasionally involve malfunction of controls resulting in fiber damage by scorching, or by mechanical abrasion of fibers in an excessively dry state.

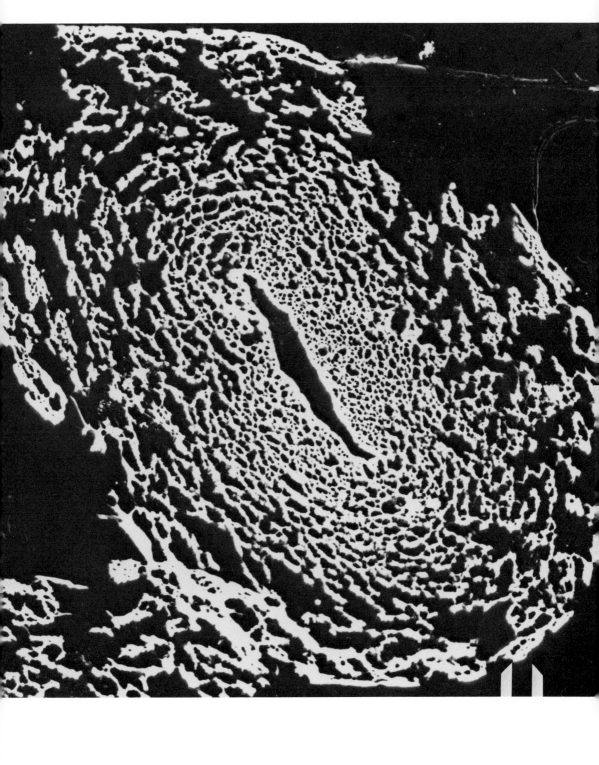

PLATE 57.

Magnified view of the edge of the heat-damaged fiber of Plate 56.

The layers are discontinuous and the microfibrils are indistinct. In oxidative damage, the cellulose becomes readily alkali soluble.

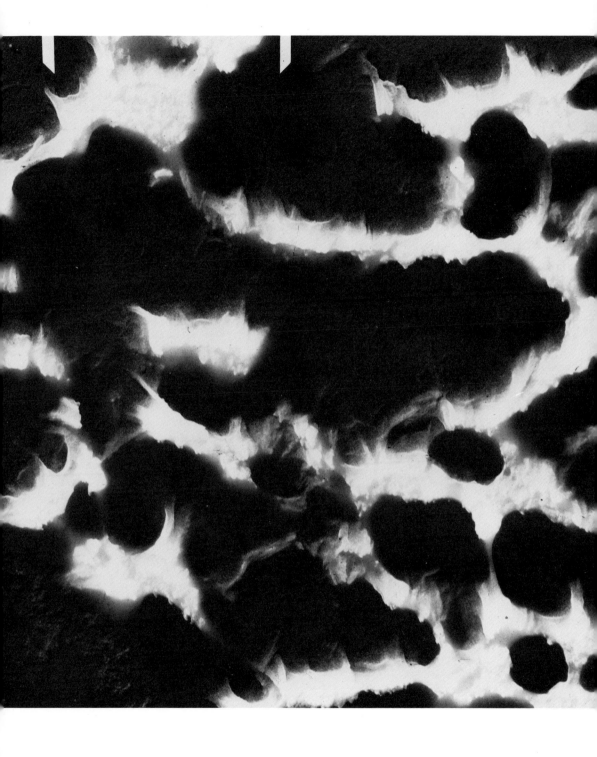

PLATE 58.

Ultrathin section of mercerized cotton fiber showing the characteristic "honeycomb" pattern after layer expansion. Compare this with the section of native cotton in Plate 46.

During mercerization, the cellulose is plasticized by swelling in a sodium hydroxide solution, and microfibrils are partially fused. Subsequently, on removal of hydroxide by washing and acid-scouring, the cellulose dries into a somewhat altered pattern, as reflected in this photograph.

Mercerized fibers are more absorptive and more lustrous than those of native cotton.

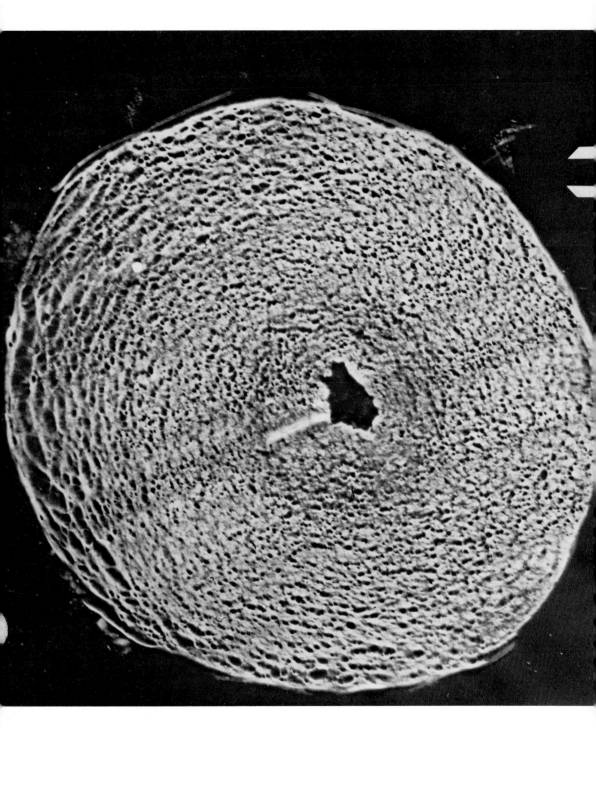

PLATE 59.

Enlargement of the outer edge of the fiber section shown in Plate 58.

The structure appears extremely different from that of the native cotton fiber of Plate 47, although both have responded to the water-methacrylate expansion method.

PLATE 60.

Layer expansion pattern in ultrathin section of mercerized cotton fiber treated with 1,10-dichlorodecane.

Although somewhat different from the fiber of Plate 58, this specimen still shows the characteristic network of mercerized cotton after layer expansion.

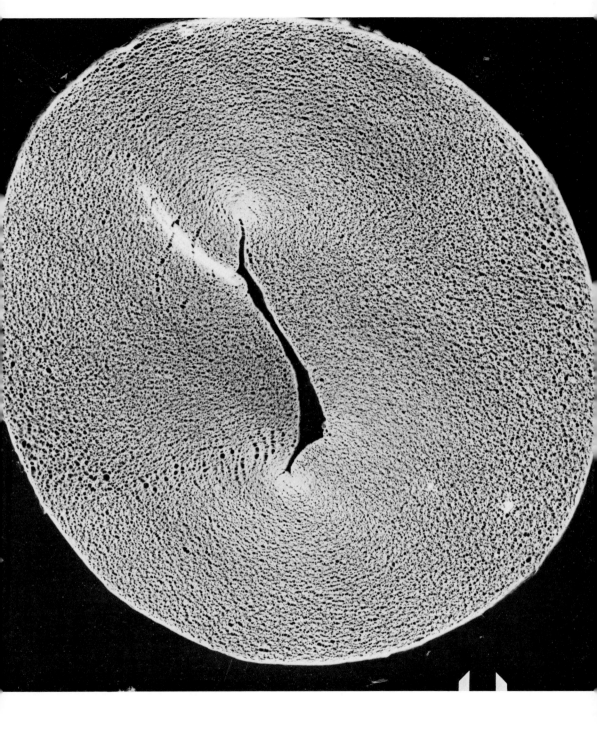

PLATE 61.

Magnified edge of the section of Plate 60.

The honeycomb structure seen here appears to be even more open than that of the mercerized cotton shown in Plate 59.

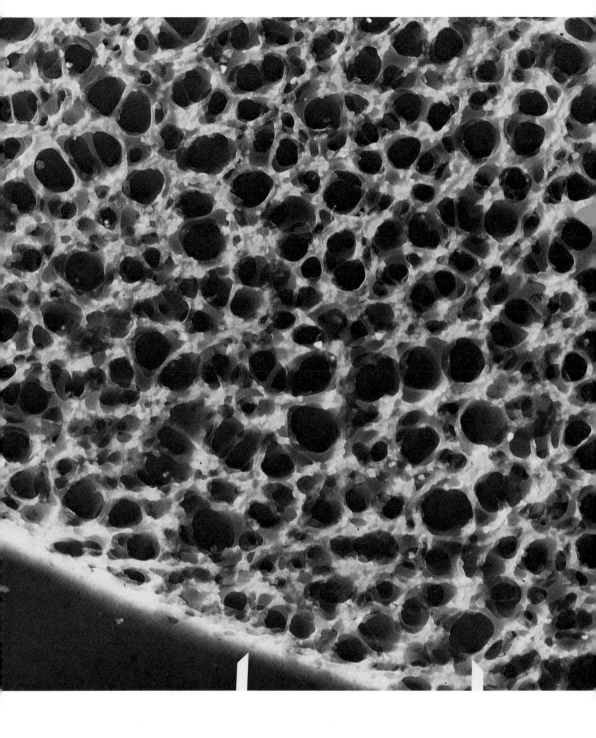

PLATE 62.

Ultrathin section of cotton fiber treated first with sodium hydroxide, then with butadiene diepoxide.

    After being subjected to the layer expansion technique, this fiber appears to be completely solid with no tendency to layer. Evidence of good crosslinking is seen here.
    Compare with Plate 48 and Plate 50.

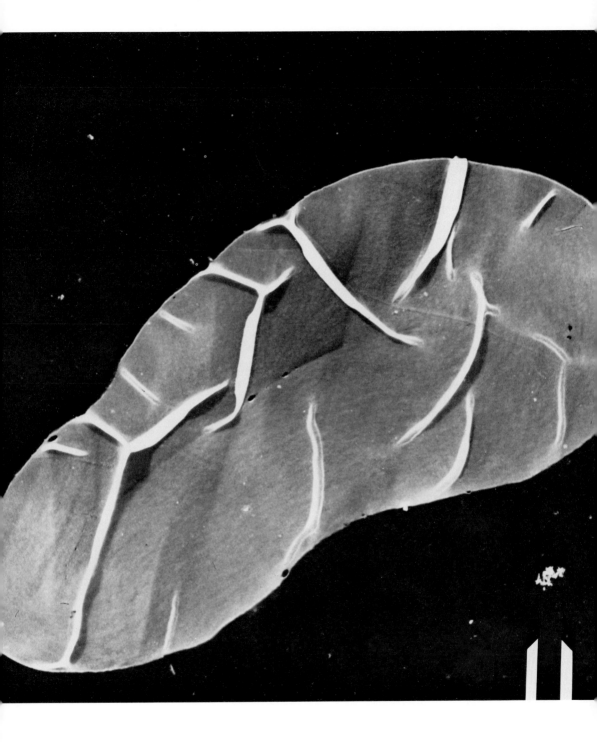

PLATE 63.

Enlarged portion of the fiber section of Plate 62.

The texture of the secondary wall is solid and nongranular except for small spheres on the surface, probably residual from the treatment but foreign to the cotton.

PLATE 64.

Carbon replica of native cotton fiber in which surface rugosities characteristic of untreated cotton are reproduced.

A two-stage replication technique reveals fiber surface structure.

The corrugated topography here recorded is produced by the wrinkling of the primary wall during the shrinkage of the fiber in its initial drying out. The noncellulosic constituents of the primary wall prevent its shrinking as much as does the more nearly pure cellulose of the main body of the fiber. The rugosities thus formed persist to some degree through most finishing treatments, such as scouring and bleaching. They are largely removed in mercerization involving high tension.

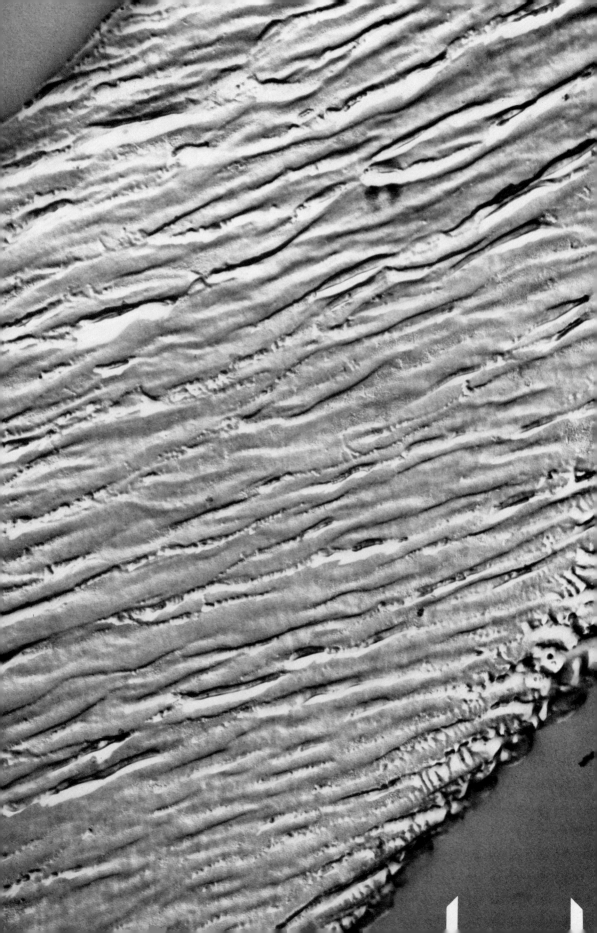

PLATE 65.

Surface replica of scoured cotton showing grooves apparently less deep than those seen in the native cotton fiber of Plate 64.

Extraction with 95% ethanol followed by scouring (pressure boil) with 1% sodium hydroxide removes much of the waxes, fats, pectins, and other primary wall impurities. However, microfibrils are not completely freed until several hours of additional boiling in ethanolamine.

The conventional commercial scour in dilute caustic (without pre-extraction in alcohol) removes only about two thirds of the noncellulosic material in cotton, most of which is contained in the primary wall.

PLATE 66.

Carbon replica of cotton fiber scoured and then bleached to show surface characteristics and markings.

The fibrillar network of the primary wall skeleton is beginning to be uncovered, but only near the edge of this fiber is it fully revealed.

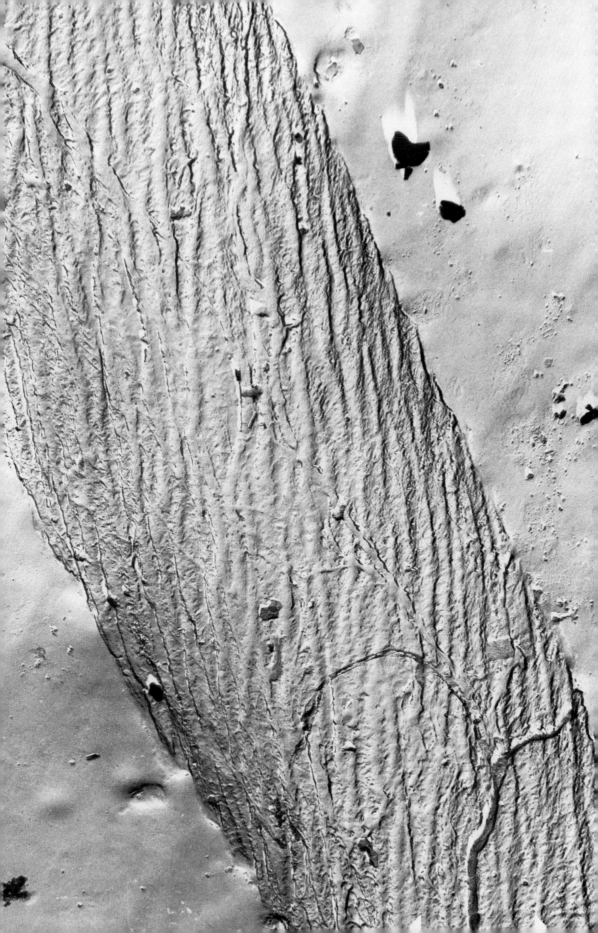

PLATE 67.

Replica of the surface of a slack-mercerized cotton fiber.

Compared to the deep grooves of a native cotton fiber (Plate 64), this replica gives the impression of relative smoothness. Mercerization is achieved by swelling the fiber in a strong solution of sodium hydroxide. If the fiber is stretched while in the swollen plasticized state, most of the wrinkles disappear and the orientation of residual ones is more nearly parallel with the fiber axis.

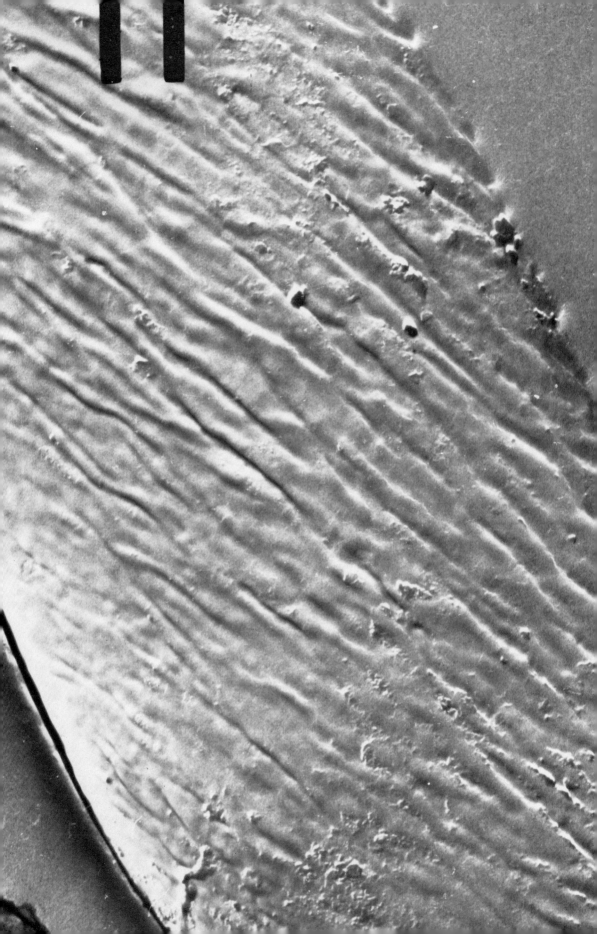

PLATE 68.

Carbon replica of cotton fiber damaged by repeated abrasion in the dry state.

The surface smoothness, the appearance of deep cracks, and the development of loose particles are typical of dry abrasion of an untreated fiber.

The carbon replica technique is valuable in demonstrating the effects of various types of damage.

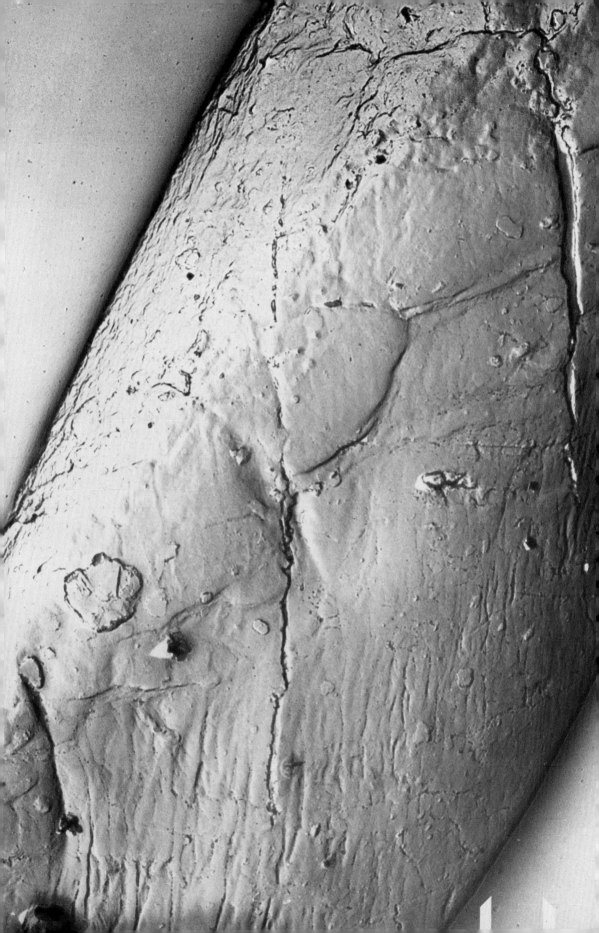

PLATE 69.

Carbon replica of fiber from untreated cotton fabric subjected to seven commercial launderings.

Fibrillation and peeling of the surface are characteristic of wet abrasion damage on cotton. Note that the primary wall has been peeled off in much of the area and the paralleled fibrillar texture of the secondary wall is revealed.

Compare this wet-abraded surface with the typical dry-abraded surface of Plate 68.

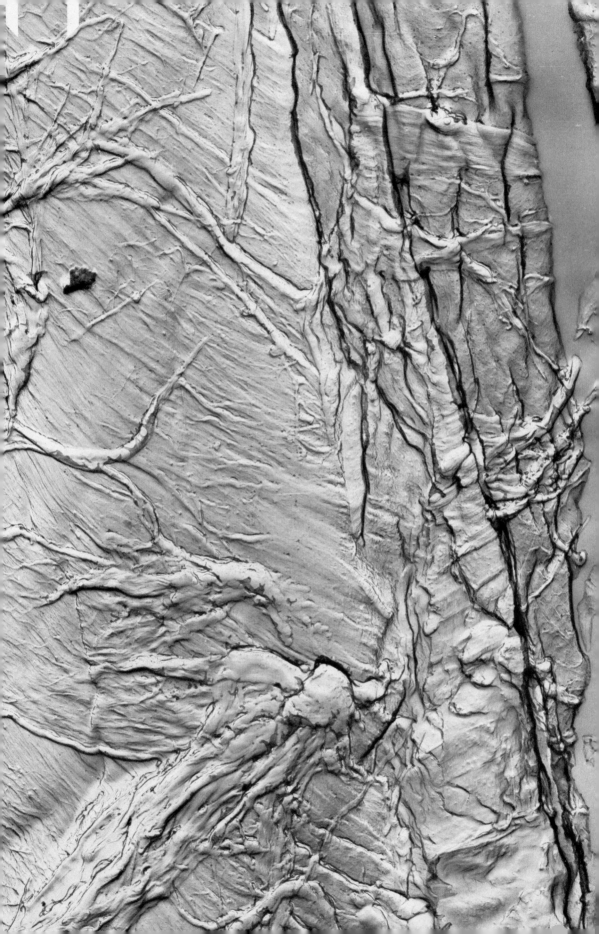

PLATE 70.

Replica of the surface of a fiber from a wrinkle-resistant fabric.

Depending on conditions of treatment, crosslinked fibers often have relatively smooth surfaces free of outstanding characteristics. This is especially true of fibers premercerized before crosslinking, or of fibers from a treatment involving deposition of a resin.

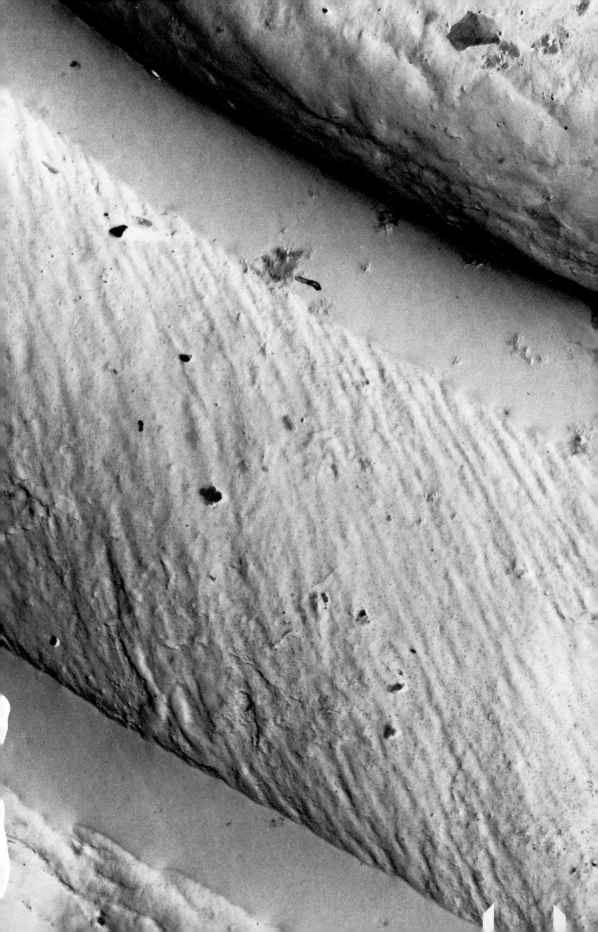

PLATE 71.

Surface replica of a crosslinked cotton fiber flex-abraded to rupture in a laboratory flex tester.

The primary wall has been scrubbed off and the secondary wall (the main body of the fiber) shows peeling and tearing, with several cracks.

3

SCANNING ELECTRON MICROSCOPY

Although the concept of the scanning electron microscope (SEM) dates back nearly as far as does the transmission electron microscope (TEM), this instrument has only been available commercially since 1965. These two differ fundamentally in principle but complement each other in ultimate results obtained.

The scanning technique presents a three-dimensional image more closely resembling that seen in the light microscope than in the TEM. Resolution with present scanning microscopes falls between that of the light and that of the transmission electron microscope. Since it permits the study of surface structure with a very large depth of field, it is ideally suited to the study of relatively thick structures in which topographical information is sought.

The SEM has added a new facet to the study of the gross morphology of cotton. Abrasion damage, surface coatings, and markings on fibers are presented in a different aspect than those yielded by the other microscopes.

Sample preparation for the SEM is generally quite straightforward. Direct examination without replication is possible in most cases. Samples may vary in size, but those measuring up to 1 cm² can be easily accommodated simply by depositing on, or cementing to, the metal specimen holder. Larger samples can often be accommodated by use of a special holder. If the sample is nonconductive, it must be treated to impart conductivity. This may be accomplished by placing the sample in a vacuum evaporator and coating it with a continuous micro–layer of a heavy metal, such as gold. In addition to providing electrical conductivity, the gold layer also controls, to some extent, the scattering cross section and the resolution obtained.

Specimens are not "shadowed" in this process. Contrast in the SEM is obtained by variations in surface topography rather than by metal thickness, as in the case with the TEM. Samples are usually mounted in the microscope with their planes at an angle to the electron beam; the most common angle used is 45°. Since the direction of the beam is also the apparent direction of viewing, pictures appear as if the sample were viewed at an angle.

PLATES

PLATE 72.

Single cotton fiber.

It is difficult to view an entire cotton fiber microscopically because of the extreme of its length-diameter ratio. For this reason, the many structural changes which occur in the fiber from its base to its tip are not completely appreciated. By coiling a single fiber around a needle and pressing this coil onto a tacky substrate, the entire fiber may be viewed at once in the SEM.

Magnification x120

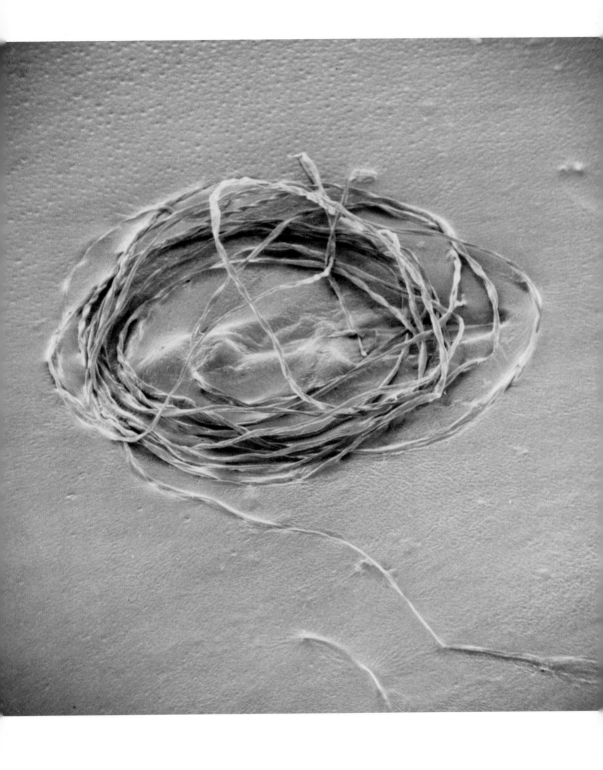

PLATE 73.

Immature fiber.

A cotton fiber reaches maturity in approximately 60 days. During this time, the fiber thickness increases to an average of $15-\mu$. When an immature fiber collapses on drying, it forms a much thinner ribbon than does a mature fiber.

Magnification x1400

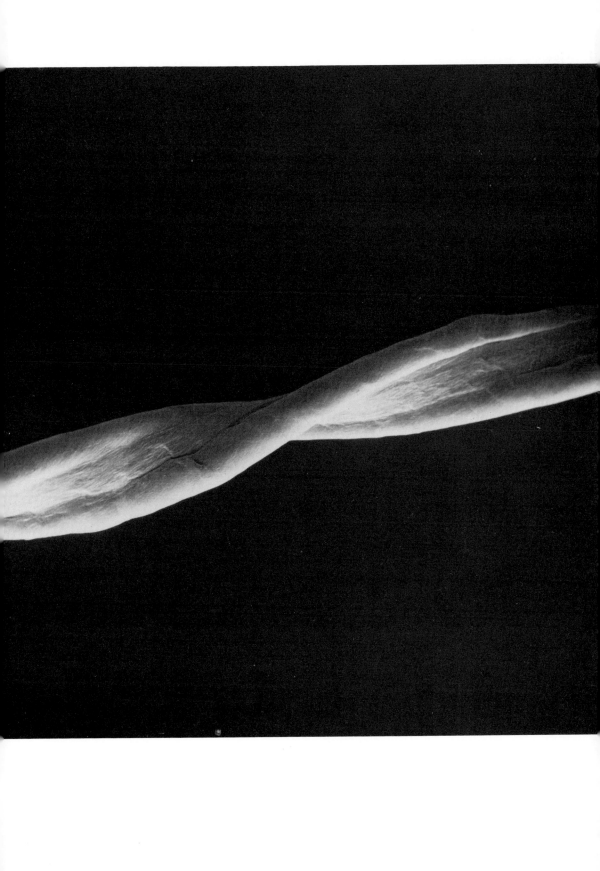

PLATE 74.

Convolutions in mature cotton fibers.

The growing cotton fiber is a hollow tube which collapses on drying with the consequent formation of twists or convolutions. This collapsing and twisting causes opposite fiber walls to take on convex-concave formations. As the twist progresses down the fiber, the convex wall often becomes the concave wall. Thus, the fiber has no true "front and back," but is made up of a series of continuously changing structural zones. Plate 74(a) shows a convolution and the rounded convex outer fiber wall. Plate 74(b) shows, in the same fiber, a zone in which the curvature has been reversed and the outside wall of (a) has become the inside wall. In Plate 74(a) the two outer edges of the fiber, formed when the fiber flattened on drying, have folded together and give the appearance of being two separate fibers. The very pronounced wrinkling of the primary wall is also evident.

Magnification x2000

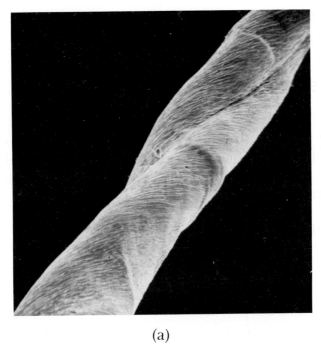

(a)

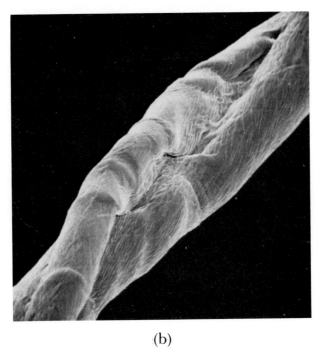

(b)

PLATE 75.

Compression marks.

In many areas of the cotton fiber, ridges or compression marks are evident on the fiber surface. These ridges are believed to be caused by folds in the secondary layers which are covered over by the primary wall.

Magnification x4000

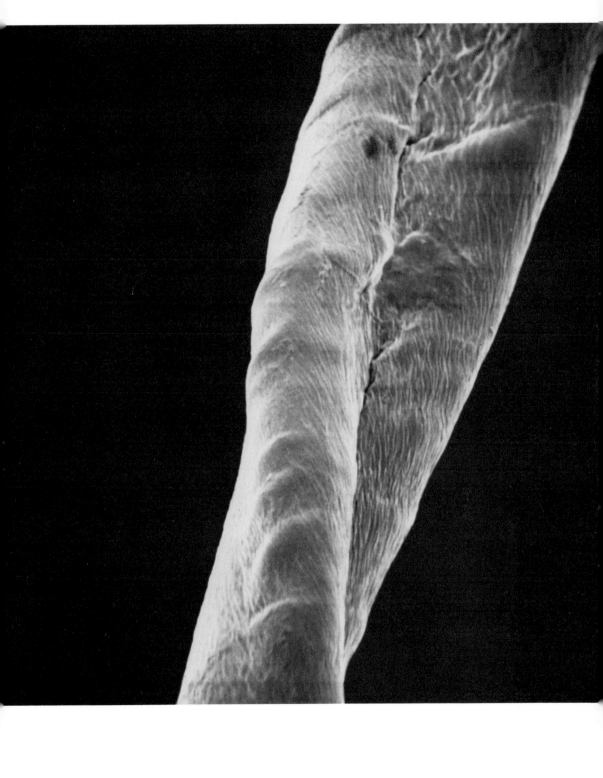

PLATE 76.

Reversal.

At intervals along the cotton fiber, reversals, or changes in the direction of the helix angle of the fiber occur. Reversals can be seen in the light microscope by the use of polarized light. In the scanning electron microscope, they are detected by the change in direction of the wrinkles in the primary wall.

Magnification x4000

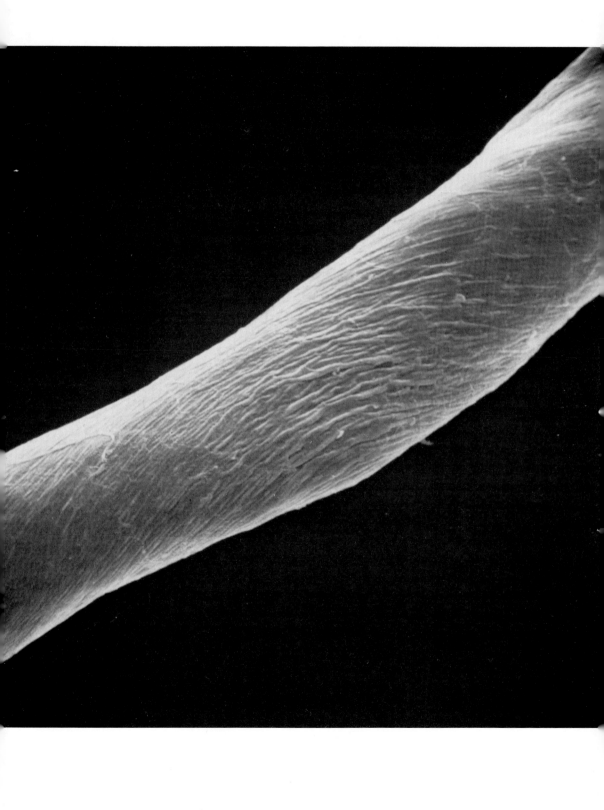

PLATE 77.

Mercerized fiber.

The surface of a well mercerized fiber takes on an appearance similar to that of many synthetic filaments. The mercerized fiber shown is almost cylindrical and has a relatively smooth surface. The primary wall wrinkles characteristic of the surfaces of untreated cotton fibers have been smoothed by the internal swelling of the fiber.

Magnification x4000

PLATE 78.

Fungus damage.

Cotton, as well as other natural cellulose fibers, is susceptible to damage by microorganisms. These organisms are present in soil, air, and even on the surface of the fibers themselves. Under proper conditions of temperature and humidity, the organisms become active and can cause drastic structural damage in cotton fibers. Plate 78(a) indicates that initial damage occurs as cracks or separation of fibrils occur in the secondary wall which proceeds in the direction of the spiral angle and breaks through the primary wall. It is also noted in Plate 78(a) that the cracks are propagated in opposite directions on opposite sides of a reversal. Plate 78(b) shows more extensive damage as chunks of the fiber have been broken away.

Magnification x1000

(a)

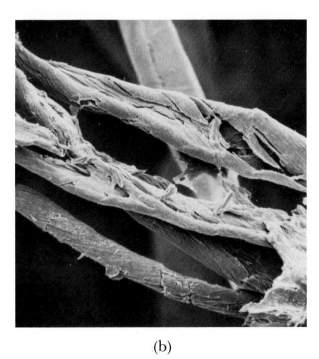

(b)

PLATE 79.

NaOH swelled fungus damage.

Damage in cotton fibers is often undetected even with the aid of a microscope. In such cases, swelling in NaOH exaggerates any cracks or breaks present and makes them more visible. A cotton fiber is shown which has been damaged by microorganisms, then swelled in NaOH. The segmented appearance of the fiber is typical of heat or fungus-damaged fibers which are post-swelled in caustic.

Magnification x5600

PLATE 80.

Grey fiber, wet-abraded.

Abrasion damage is a major problem in the utilization of cotton fibers in textiles. Abrasion reduces both tear and breaking strength of cotton fabrics, as well as reducing their esthetic appeal. Wet and dry abrasion produce different types of damage in the cotton fiber. When the fiber is wet, it swells to an extent great enough for fibrils to be rather easily separated by physical forces. Illustrated is a native, untreated fiber from which the primary wall and winding layer have been removed after abrasion in a washing machine. Extensive fibrillation is noted in the primary wall, and the portions torn from the fiber appear stringy.

Magnification x4000

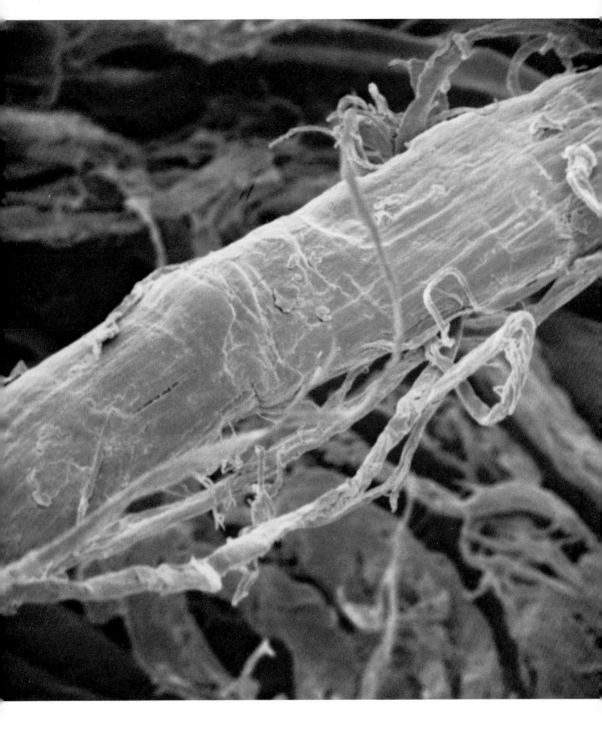

PLATE 81.

Grey fiber, wet-abraded end.

Broken ends develop in wet-abraded native fibers where they have been shredded and pulled apart. The ends appear as frayed fibrillar bundles.

Magnification x4000

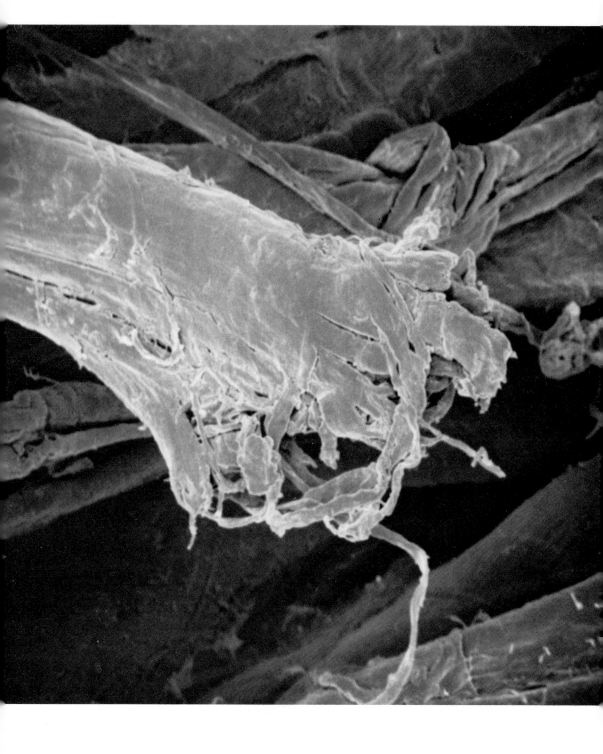

PLATE 82.

Crosslinked fiber, wet-abraded.

Wet abrasion of crosslinked fibers is very similar to that of untreated fibers. Damaged fibers splinter and peel, but the peeling is more of fibrillar sheets than of bundles of individual fibrils, since fibrils are bonded together in the crosslinking reaction. Plate 82 shows the beginning stage of breakdown in a fiber cross-linked with tris(1-aziridinyl)phosphine oxide-dimethyloldihydroxyethyleneurea (APO-Permafresh 183). The outer layers have been split and peeled back, revealing the fiber body beneath.

Magnification x4800

PLATE 83.

Crosslinked fiber, wet-abraded.

More extensive damage with the separation of sheets of fibrils is seen in the fiber, crosslinked by the same procedure.

Magnification x4800

PLATE 84.

Grey fiber, dry-abraded.

Dry abrasion produces a different type of damage than wet abrasion. Fiber surfaces become smoother and have no evidence of fibrillar texture. Major damage consists of diagonal or horizontal cracks across and at the edge of fibers. Such damage is shown in the grey fiber illustrated, with the crack following the direction of the winding layers. This abrasion was brought about in a clothes dryer. As moisture is driven out of the fabric in the dryer, the fibers tend to collapse, and cellulose-to-cellulose hydrogen bonds become stronger. Fibrillar units are more tightly bound, increasing fiber stiffness. Embrittlement of the fiber, caused by the reduced possibility of internal fibrillar slippage, results in the cracking or breaking of the fiber on stress. Crosslinked fibers abrade in the dryer in much the same way.

Magnification x4000

PLATE 85.

Grey fiber end, dryer-abraded.

This cotton fiber shows the type of break which can occur when dry fabrics are abraded in a clothes dryer. The smooth end shows no fibrillation. Often the ends are so straight and smooth that they appear cut instead of broken.

Magnification x4000

PLATE 86.

Scoured yarn cross section.

The maturity of cotton fibers is readily determined by viewing their cross sections. More mature fibers have oval or bean-shaped sections, while immature fiber cross sections are thin walled and distorted in shape. Most cotton yarns contain fibers of various degrees of maturity, as illustrated.

Magnification x2000

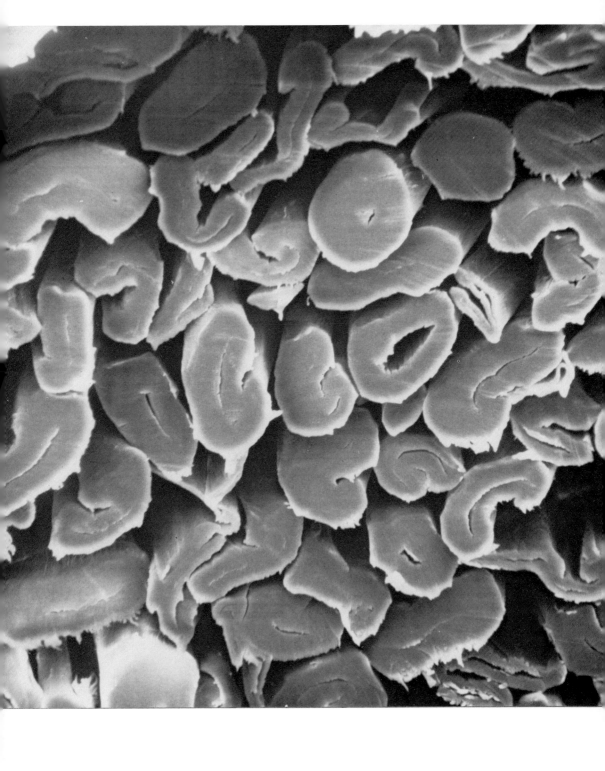

PLATE 87.

Mercerized yarn cross section.

Mercerization has a very drastic effect on the fiber cross section. As the fibers swell and become rounded, their cross-sectional areas also become rounded, and in some cases almost circular. Often the swelling closes the lumen completely.

Magnification x2000

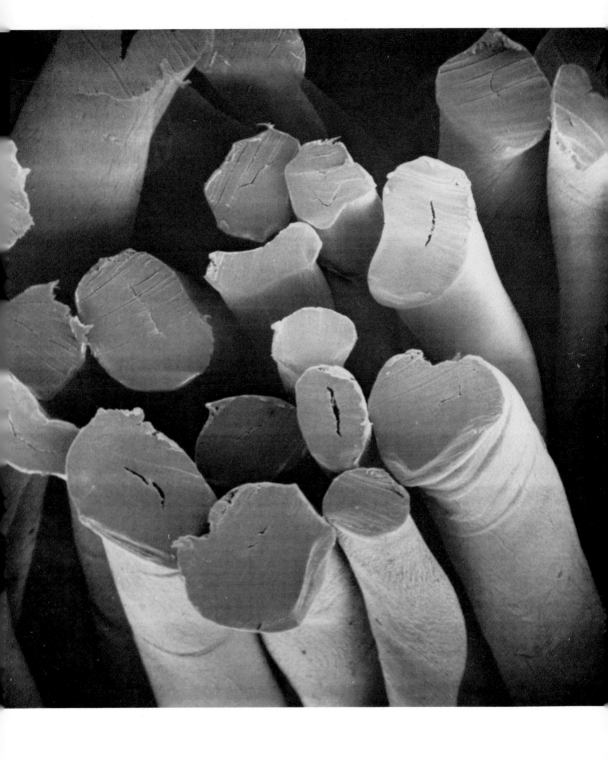

PLATE 88.

Flame-resistant yarn cross section.

A cross section of a yarn from a burned flame-resistant fabric shows that the fiber wall thickness has been reduced by as much as one-third, as compared to Plate 86, and the fibers have become hollow tubes fused together when the polymer melted on application of heat.

Magnification x2000

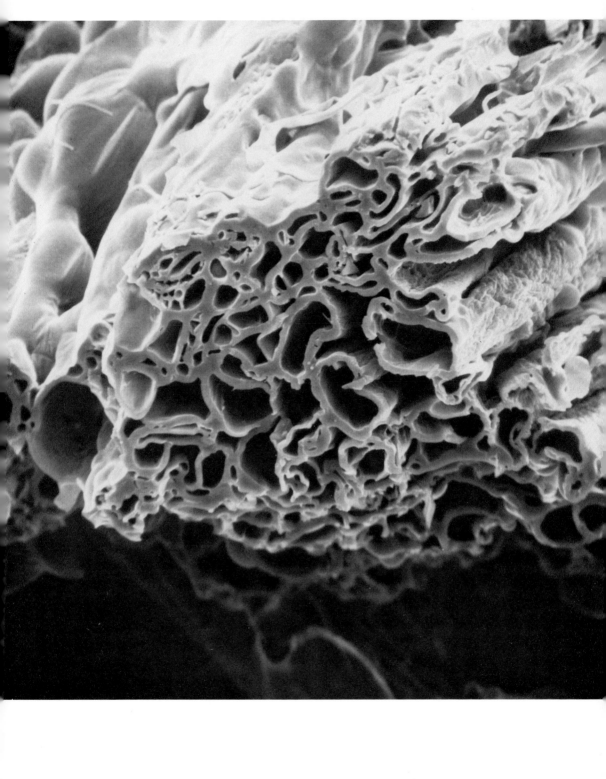

PLATE 89.

Scoured fabric.

The structure of a scoured 80^2 printcloth is shown. The SEM shows (Plate 89a and b) the relationship of the yarns to each other as well as fiber-fiber relationships within each yarn.

Plate 89(c) shows, at high magnification, the way a group of fibers lie upon one another at the surface of a yarn.

Magnifications x50, x100, and x1000

(a)

(b)

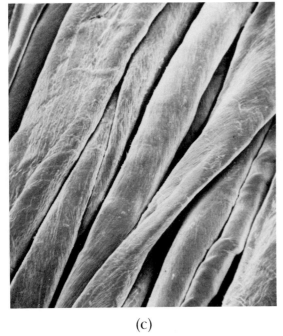

(c)

PLATE 90.

Mercerized fabric.

Mercerization is the process by which cotton is finished by treatment in strong caustic soda. Conditions of mercerization vary widely and seldom is complete mercerization, or conversion to Cellulose II, achieved in commercial mercerization processes. Without the use of a microscope, little difference can be seen between a nonmercerized and a mercerized fabric, except that the mercerized fabric has an enhanced sheen. In the SEM, other changes are obvious. A slack-mercerized fabric (Plate 90a and b), shows shrinkage and a closed-up yarn structure. A high magnification of fibers (Plate 90c) in the yarn shows that they are more rounded and have smoother surfaces.

Magnifications x30, x100, and x1000

(a)

(b)

(c)

PLATE 91.

Fibers from acrylic-coated fabric.

Various polymers are sometimes used in fiber finishing. Depending on the purpose, the polymer is either deposited inside the fiber or on its surface. In addition to coating fiber surfaces, the polymer can form bridges which bond fiber to fiber, and even yarn to yarn. Illustrated is the surface appearance with polymer bridges of a fabric treated with 12% of an acrylic polymer. Depending on the nature of the polymer and the type of fabric to which it is applied, some bridging can be seen with polymer contents as low as 0.2%.

Magnification x4000

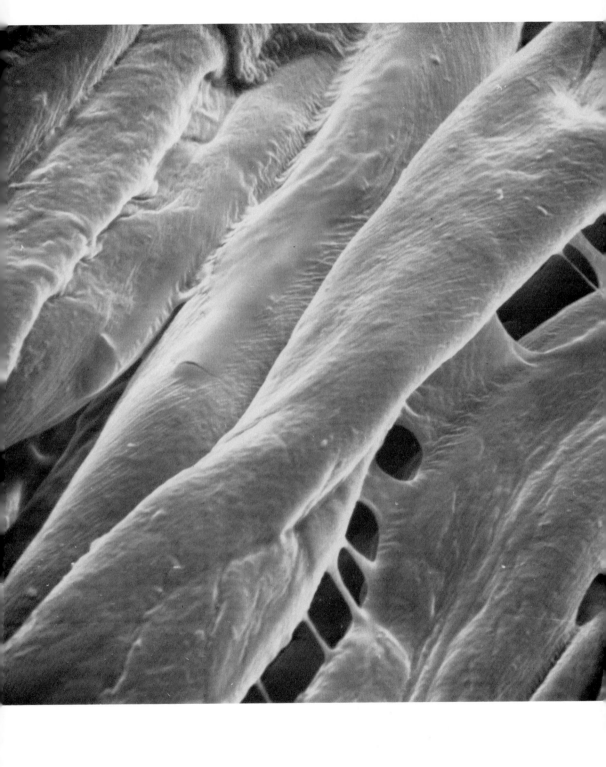

PLATE 92.

Fibers from polymer-coated fabric.

Polymer treatments do not always produce an even distribution from fiber to fiber. A fiber is shown which is covered with clusters of the deposited polymer, while the fiber in the background has a clean surface.

Magnification x4000

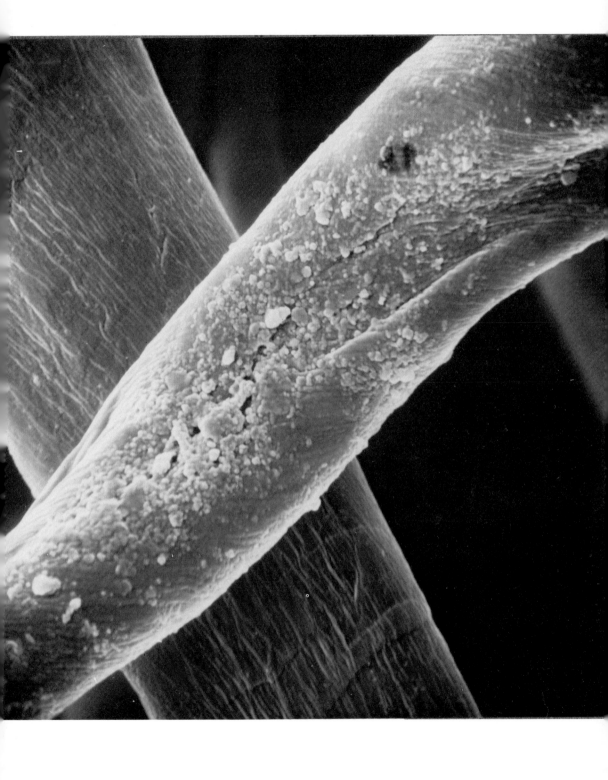

PLATE 93.

THPOH-NH₃-treated fabric showing deposit.

Flameresistance is imparted to cotton fabrics by treating them with phosphorus- and nitrogen-containing polymers. While these polymers are intended to deposit within the fiber, much of the material remains on the fiber surface. Fibers are shown from the surface of a fabric treated with a 40% solution of solids formed from the reaction of tetrakis(hydroxymethyl)phosphoniumchloride with sodium hydroxide. The tetrakis(hydroxymethyl)phosphonium hydroxide (THPOH) thus formed is padded onto the fabric which is then treated with ammonia to insolubilized P-methylol polymers. It is obvious from the illustration that much of the polymer has remained on the fiber surface.

Magnification x4000

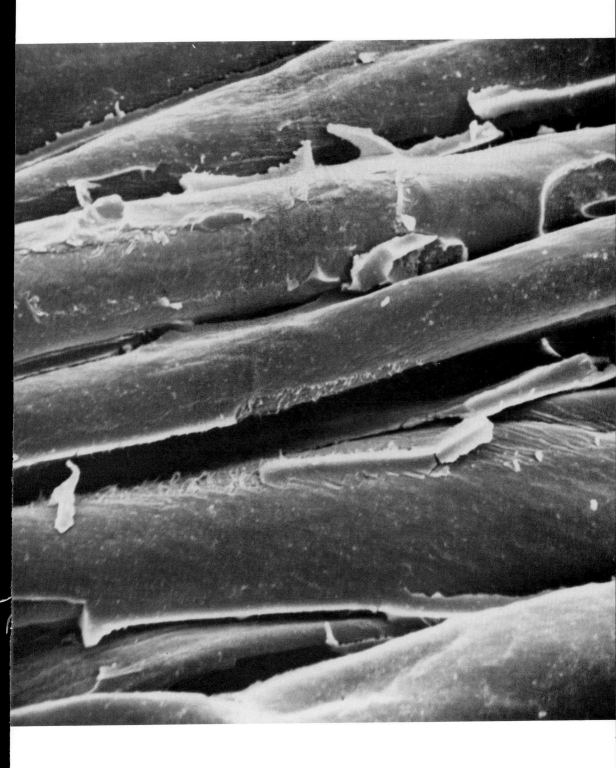

PLATE 94.

THPOH-NH₃-treated fabric, burned.

When the THPOH-NH₃-treated fabric is burned it retains its fabric structure although the char formed is very brittle. At low magnification, the fabric appears little different from an unburned fabric (compare Plate 89).

Magnification x200

PLATE 95.

Burned fibers from THPOH-NH₃-treated fabric.

At higher magnification, the different structure of the burned fiber becomes obvious. The polymer is fused, coating the fiber surface and forming bubbles.

Magnification x4000

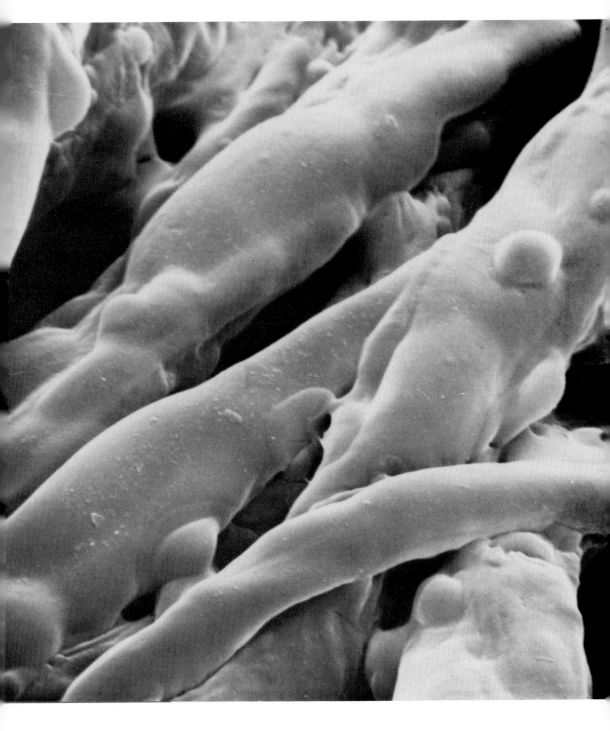

PLATE 96.

Fiber from APO-THPC-treated fabric, burned.

A similar flame-retardant treatment consists of reaction of the cellulose fibers with a mixture of tris(1-aziridinyl)phosphine oxide (APO), and tetrakis(hydroxymethyl)phosphonium chloride (THPC). When reacted in the roving form, little polymer is built up on the fiber surface. However, when it is burned the fiber has an appearance similar to that of the THPOH-NH$_3$-treated fiber. Even though the fiber surface is greatly distorted by bubble formation, its original wrinkled appearance is still evident in certain areas.

Magnification x4000

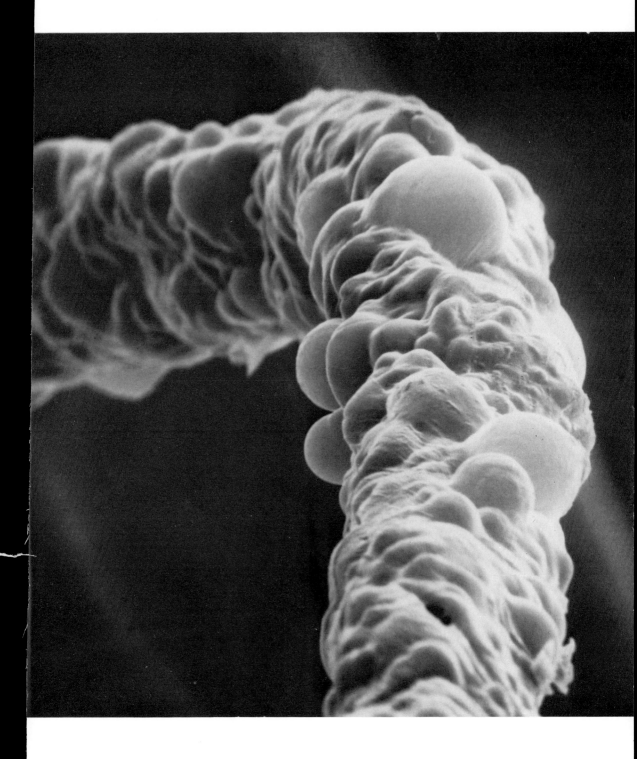

PLATE 97.

THPOH-NH₃-treated fabric, laundered.

One requirement for a fire-retardant treatment is that it be durable. Durability may be tested by a series of laundry cycles. Most fabrics with surface treatments progressively lose their deposits through a laundry series. Initial removal of the deposit is generally evident by the fifth laundering. The surface of a fabric treated with THPOH-NH₃ is shown after 20 launderings.'Most of the surface deposit has been removed, and laundry abrasion has torn away the surface of many fibers as well.

Magnification x2000

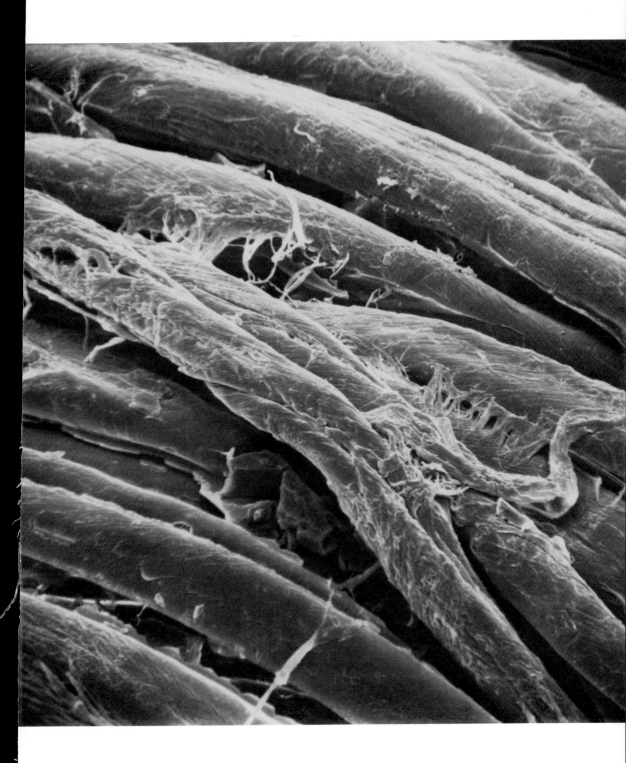